VIDEOJOURNALISM

THE DEFINITIVE GUIDE TO MULTI-SKILLED TELEVISION PRODUCTION

VIDEOJOURNALISM

THE DEFINITIVE GUIDE TO MULTI-SKILLED TELEVISION PRODUCTION

Richard Griffiths

Focal Press
An imprint of Butterworth-Heinemann
Linacre House, Jordan Hill, Oxford OX2 8DP
A division of Reed Educational and Professional Publishing Ltd

A member of the Reed Elsevier plc group

OXFORD BOSTON JOHANNESBURG
MELBOURNE NEW DELHI SINGAPORE

First published 1998

British Library Cataloguing in Publication Data
A catalogue record for this book is available from the British Library

Library of Congress Cataloguing in Publication Data
A catalogue record for this book is available from the Library of Congress

ISBN 0 240 51508 0

Composition by Genesis Typesetting, Laser Quay, Rochester, Kent
Printed and bound in Great Britain

Contents

Preface

A station I worked at years ago had a problem. We wanted to produce lively news programmes with plenty of location interviews but couldn't afford £800 a day for a crew. Then I discovered that a local facilities house was hiring out one of the brand new Hi-8 camcorders that had just emerged onto the market for less than £100 a day. This was more like it. I brought it up at the weekly editorial meeting. Why couldn't we, I suggested enthusiastically, hire the Hi-8 camcorder for a day and knock off as many interviews as we could. As news editor I would set the interviews up and conduct them myself and the channel's in-house director could shoot them. A hush descended on the room. The director bristled and glared. He drew himself up to full height. 'But . . . I'm a director,' he growled, as though I'd asked him to clean out the bins.

I looked around for support and suddenly became aware of the general level of animosity created by this horrendous breach of protocol. God forbid that a television director should actually get his hands dirty by operating a camcorder. In my ignorance I backed down completely and the programmes never got made. That, it appeared, was preferable to the solution I had suggested.

In the early part of 1994 the publishers of the *Evening Standard*, Associated Newspapers, advertised in *The Guardian* for 24 videojournalists. It was the first time this job title had been used in the British broadcasting industry. We had heard, of course, about videojournalism in America, where reporters were sent out on their own with a camcorder to shoot the pictures, direct the sequences, conduct the interviews, record the pieces-to-camera, write the scripts, and package the features. But Britain was the last bastion of quality broadcasting in the world so how could such a concept be imposed on our television professionals? The industry was appalled – and yet intrigued. It watched and waited. The 24 videojournalists were needed to staff Associated Newspaper's brand new cable news station for London, Channel One. Full training would be given. It was an offer I couldn't refuse.

The second of the interviews was with an astute American called Michael Rosenblum. He sent me out with a camcorder and gave me half an hour to shoot a story. When I returned, he spun through the rushes and offered me the job. Videojournalism is *that* practical. On the first day of his inspiring and intensive training course, he told us this: 'Any idiot can make television.' And when we had finished, we were indeed producing network-quality news features completely on our own. It was a liberating experience. Perhaps one of us had picked up a camcorder before. Maybe a couple had worked in some capacity in television before. But all of us were experienced journalists with print or radio training behind us and we all wanted to learn something new.

Channel One itself boasted that it had spent £10,000 training each and every one of us. The reason they did it, of course, was because in the long run it was cheap – Channel One was one of the first of the many new channels that was forced to produce their programmes very cheaply simply because there weren't enough subscription-paying viewers to justify charging much for advertising.

Since then, videojournalism has gone through a learning curve. It is a superb method of production for story-led news, features and documentaries. The intimacy of one person and a small camcorder allows the videojournalist to get closer to the subject than an intrusive crew could ever manage. But it is really no good for live broadcasts (who changes the iris when the sun comes out from behind the cloud?), court reports (either you get the story inside or the pictures outside – you can't get both) or presenter-led productions (though a videojournalist can quite happily do all the other jobs, as this book will show).

Technology has also come to our aid. In America, videojournalism was pioneered using cheap Hi-8 camcorders that in the right hands could produce just about broadcast-quality pictures on NTSC tape (it's not known as Never The Same Colour for nothing). To get the same results on PAL tape required a really skilled camera operator, the right lighting conditions and the ability to edit directly from the original Hi-8 rushes. It was all too much and Channel One chose instead to equip itself with bottom of the range Beta SP camcorders costing four times as much as Hi-8 camcorders. The digital DVC camcorders, though, are another matter. Any trained professional can produce broadcast-quality pictures using one and they will be the driving force behind the expansion of videojournalism in the PAL world.

As more low-budget channels are launched and conventional broadcasters are forced to cut costs, so the demand for high-quality cost-efficient programming will soar. Multi-skilled television production techniques like videojournalism will become commonplace. I agree with Michael Rosenblum that any idiot can make television (don't we all just know it) – but it takes training and skill to make *good* television.

This book helps you get there.

Introduction

One person and a camcorder. It may seem inconceivable that this combination could produce as good, if not better, broadcast television as a whole production team but stick with me and we'll prove it to be true. You will learn how to:

- combine technical and editorial skills into one job function, and you will wonder how they were ever kept rigidly separate;
- choose stories that are inherently suited to television – because there's nobody else to blame when you end up taking endless stock shots of someone typing at a computer or (heaven forbid) making a cup of tea;
- shoot your stories to edit – that is, to work out exactly how you are going to put your piece together as you are getting it;
- walk into a situation where things are happening and have your first sequence in the can in the time it takes a crew to unpack their tripod.

But, and it's a big but, you have to make a commitment to follow the instructions to the letter. We have to be strict about this.

As you will already have realised, videojournalism demands a lot from the videojournalist. It requires editorial and technical proficiency across a whole host of different areas. You may feel that you are already up to speed on certain aspects of the job – and that's great, because it means you have a slight head start – but videojournalism is all about meshing these different skills into one job function, and to do that you have to go right back to basics. I don't want to see people trashing multi-skilled production as a concept because they have skip-read passages of this book and haven't gone out and done all the exercises. You can't beat good training.

So what exactly is videojournalism? Quite simply it's a cheap way to make high-quality story-led television. One person becomes a camera operator, sound operator, director, production assistant, scriptwriter, producer and presenter. Videojournalists don't edit their own pieces, even though they may be proficient editors. Anyone involved in the media knows that an objective eye is fundamentally important to the production of a polished finished product and the same is true for videojournalism. Script editors and video editors are the safety net for videojournalists.

Videojournalism can be used to make news, features and documentaries that are driven by the stories and the subjects. I once made a very simple and effective two-minute item after I had been taken round a house that had been an inferno just minutes before: the glowing pictures were great and the fireman's narration chilling. Conversely, videojournalism can be used to make multi-part series that go into great depth about subjects such as motoring, travel, fashion and food.

The point is that the operative word in videojournalism is journalism and, while the videojournalist narrates the package and will include a piece to camera, the technique is not suitable for presenter-led television simply because there's nobody else to hold the camera – though you'd be surprised how inventive some videojournalists can be to make sure they get their faces on screen as much as possible!

Part of the videojournalist's job is to be as up to date as possible on the latest technological breakthroughs in the broadcast media because, frankly, these make their job easier. Videojournalism first came about because the equipment needed to make broadcast-quality television became light enough for one person to carry and operate. And the advances go on.

But whether you are using a shoulder-mounted Beta SP camcorder or a hand-held gizmo the size of a cigarette box that stores digital pictures on computer chips, the techniques of videojournalism remain the same – as does your degree of technical proficiency. Camera manufacturers have yet to devise the perfect automatic white balance and nothing compares to the control you have with the focus and iris on manual. So it doesn't matter which camera you are using or what format you are recording on. When I refer to features that are common on practically all camcorders (such as zebra stripes) and you don't know what I'm going on about, all you have to do is remember the time-honoured acronym: RTFM – roughly translated as Read The Manual – and go away and do it.

The chapters in this book are split up into practical exercises and bog-standard instruction. Scattered throughout are hints on how you can do cool things such as rack focus shots. You will learn how to direct, how to write scripts, how to frame shots, how to interview people and how to put it all together. If you are using this book and following a formal course in Videojournalism, it should take you no more than a month before you are consistently able to produce broadcast-quality television. Practice and constructive criticism make perfect. So, what are you waiting for? Pick up your camera, put everything on automatic, make sure the battery is fully-charged and insert a tape. You're going out on a shoot.

1 Your first shoot

One of the worst jobs in television, apart from doing overnights at cable television stations that nobody watches, is to go out on the street and get some vox pops. This involves asking ordinary members of the public what they think about the issue of the day and hoping desperately that someone will come up with an answer that is lucid, pertinent and broadcastable.

Vox pops were pioneered by Esther Rantzen on the BBC programme *That's Life* and it may be no coincidence that her show is no longer on air. Now news and features use them wantonly to bump up the running time of packages when there is nothing left to say. Channel One ran a short-lived daily feature called *Cabbies Corner* that consisted of quotes from London black-cab drivers on whatever the news editor thought was the story of the day. Traipsing out at six in the morning to ask what cab drivers thought about, say, chocolate-flavoured beer, or the latest Royal scandal, was a thankless task and one that wouldn't have taken place if the news editor himself had been the one who had to go out and do it. The slot died a death shortly after I was sent out to ask cabbies what they thought about the possibility of Myra Hindley being let out of jail. Their responses were so venomous that the tasteless item was broadcast just once before being consigned to television heaven.

The other point about vox pops is that they rarely add anything worthwhile because most people's opinions about burning issues are transparently based on something somebody else has written in their daily newspaper (so it must be true, right?).

Approaching people in the street is a grotesquely awful job and the only pleasure that can be had is when they say something that you can edit in an entertaining way. I had a spectacular success when sent out by the breakfast station GMTV to ask what people were superstitious about on Friday the 13th. Most people, of course, didn't give a stuff and they waited until the camera was rolling to say so. But one poor man replied: 'I think it's all a load of rubbish . . . except of course I won't walk under ladders in case something falls on me.' Thus giving me the welcome opportunity to cut the quote after the word 'ladders' and turn him into a bit of a twit.

Go get them

So, your first exercise is to go out and get some vox pops. You will need to think about your question and how you can structure it to produce usable answers. One basic rule, ignored by people who don't know what they are doing, is that when the vox pops are cut together the videojournalist's voice should not be heard asking the question. Spot the difference between these two scenarios:

(1) You: 'Are you superstitious?'
Answer: 'No.'
(2) You: 'What are you superstitious about?'
Answer: 'I think it's all a load of rubbish, except of course I won't walk under ladders.'

It isn't very hard to see which sentence construction produces the desired kind of answer. You don't want simple 'yes' or 'no' answers, so 'are' and 'have' are bad ways to start your question whereas 'what', 'which' and 'why' are good.

Now you've got your question, put all the camera settings (focus, white balance, iris, sound levels) on automatic, venture out into the great wide world, point the camera at some faces and start asking complete strangers intimate details about their sex lives or something. This will get you used to operating the camera, working on your own and convincing people that you are a professional producing quality work using a camcorder. But even so, you'll get the point about doing vox pops. Horrible, isn't it?

Viewing your tape

When you come back, take a look at what you have just shot. Many things will be apparent and this is the reason you've done the exercise. For a start, the sound quality won't be good. The built-in microphones on camcorders are invariably useless for everything but recording background noise. We will address that later. Second, the picture quality will be ropey. Look at the flaring that takes place when the sun comes out from behind the cloud or a double decker bus rolls past and the automatic iris attempts to compensate. See how

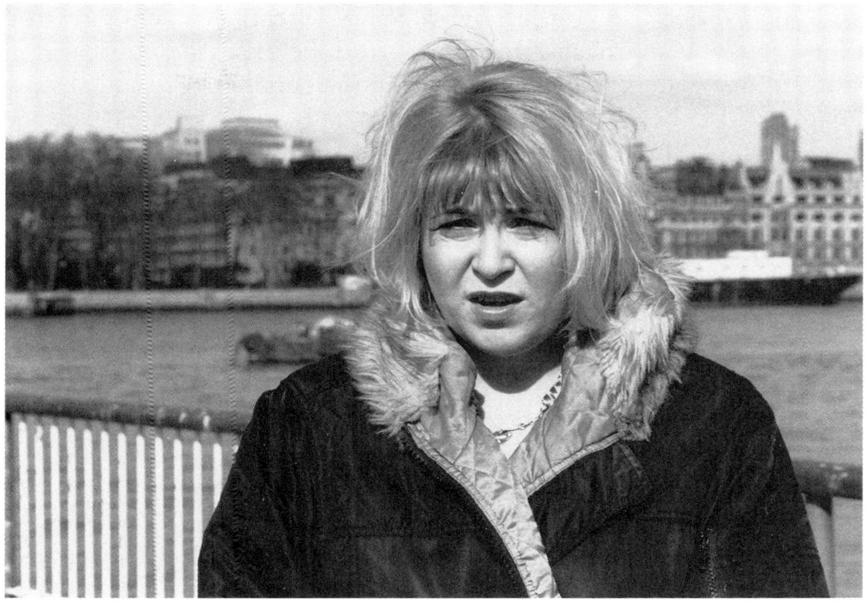

Figure 1.1 Your vox pops should look something like this

the automatic focus searches for the subject of the picture if they move their head briefly out of frame. And doesn't the colour all look a bit odd for some reason?

Next, be really critical about your shooting. How much are you wobbling the camera around? You really want to be holding it rock steady. Where is the subject's face in the frame? Ideally, you should have got a centred head and shoulders shot not views of bald patches or double chins or left ears. Figure 1.1 shows how your shot should be framed. You can see how such a simple exercise opens up a hornet's nest of issues that need to be resolved: technical, artistic and editorial.

Now do it again

Repeat the vox pop exercise keeping the camera on automatic until you are happy with your framing and the answers you are getting to your questions.

One final point – and another good reason why videojournalists shouldn't really do vox pops. To make them look really good, vox pops should be cut together so the people you are talking to look alternately camera left and camera right. This is easy if you are out with a crew because you just swap places with the camera operator every time you approach someone new. However, a videojournalist is stuck in one place behind the camera so the people's eyes will always be looking at the same point in space (usually just to the left and up a bit from the lens). It doesn't look too terrible but it's not totally satisfactory.

The main points

- Vox pops are horrible but, like political candidates, sometimes you cannot avoid them.
- You do not want answers that rely on the viewer hearing your question.
- You want more than 'yes' or 'no' answers.
- So do not start your question with 'are' or 'have', start instead with 'what', 'which' or 'why'.

2 Taking control of your camera

The technical deficiencies you suffered doing vox pops are all solvable. But don't try and deal with them all at once. Technical proficiency is something that comes with practice and time and familiarity with your camera equipment. Directors often like to take a small colour television monitor with them so they can see what shots the camera operator is getting without having to look in the evil little black and white viewfinder. Once you know your camera inside out, such palaver becomes unnecessary. You will know exactly how different shooting conditions can be compensated for and exactly how they should look in the viewfinder.

You will have seen from the vox pops why you should put as many of your camera's features on manual as possible. Automatic white balance, focus and iris never perform to broadcast standards. There are also other controls that will help you get the most from your shots. What you need to do is introduce these to your shooting regime one at a time. Play around with them. Work out what the tolerances are for producing broadcast-quality pictures and remember how those settings look in the viewfinder. Pretty soon it will all become second nature and you will be adjusting the focus and iris while reframing as though it was as basic as talking while you walk.

White balance

Think about these two instances:

(1) It's a sunny day outside and there's snow on the ground. You're indoors curled up by the fire and you decide to pop out and throw snowballs at the neighbours. You open the front door and . . . wow . . . doesn't the world look blue today? It takes your eyes a few moments to adjust.
(2) You're outside collecting logs for the fire. Into the sitting room you go where the lights are on and the fire is burning . . . hey . . .

doesn't everything have a warm red tinge? Again, your eye compensates.

The source of white light differs depending on whether you are shooting indoors with artificial light or outdoors with natural light during the day or sodium streetlights at night. Or how about when you're shooting someone who's indoors but sitting near a large window. All these situations produce different kinds of white light and the reason you are unaware of it is that your eye compensates marvellously.

Unfortunately cameras don't behave like eyes. They have no concept of what white should be, although they really should do because they base their interpretation of all other colours on it. This is why they all have a little button called white balance, which you should use as often as you feel may be necessary. All you have to do is fill the frame with white (a piece of paper or some lens caps will do the trick) and press the button until the camera's white balance indicator stops blinking. Repeat this every time you think the source of light has changed and your pictures will always come out the right colour.

Shoulder mounted camcorders often let you store a couple of white balance settings so you can easily switch between the two (say indoors and outdoors) should you ever be in such a time-critical situation. They also have a black balance button. When this is depressed, the camera shuts the iris and takes a reading. Clearly this black never changes and you should only have to do it once in a camera's life. But it's best that you know what it's for.

Iris

The iris controls the amount of light coming into the lens. When there is a lot of light the iris should be closed down and in low-light conditions it should be opened up. Again, your eye isn't much help because it automatically adjusts to the lighting conditions so you are mostly unaware of the differences (except in obvious situations such as a dark night sky or a bright white snowfield).

Camcorders will automatically adjust the iris to the optimum setting. Even when you have the camera on manual iris, there is usually a little button to press that will set it for you. The reason you must have the camera on manual iris is that you don't want the camera adjusting the iris in the middle of a shot. You really don't. It looks terrible.

Imagine for example a street scene with buses trundling past. Each time a bus cuts out some of the daylight the picture will darken. If your camera is on automatic then it will open the iris up to compensate and the picture will flare. Then the bus goes past and the daylight increases so the scene lightens, at which point the automatic iris kicks in again and causes the picture to darken before finally returning to the original setting. For broadcast this is death.

I have worked with a so-called professional cameraman who got paid the going rate every day and he invariably had the iris on automatic. Imagine the flaring and dipping when we did a walkie-talkie from the pavement into a building and down the corridor to an office past a load of windows. The results were abysmal. This is basic stuff.

So by all means use the automatic iris button to get your original setting but always keep the camera on manual. Eventually you will find that you don't even need to use the automatic button because you will know what lighting conditions look like in the viewfinder and will set the iris yourself accordingly.

Focus

There is a similar problem with using automatic focus. Shoulder-mounted camcorders don't even bother to feature automatic focus because they know professionals wouldn't use it.

Auto-focus often works effectively as long as the shot you are getting is a static scene. But as soon as something moves, the focusing mechanism will probably kick in and off it will go searching near and far for the new setting, ruining your shot in the process.

There are a number of ways to focus manually depending on the type of camera you are using. With shoulder-mounted camcorders, you should zoom into the subject, focus until it is sharply pin-pointed in the viewfinder (make sure the viewfinder is adjusted to your eye or all your shots will look soft) and then zoom out and frame accordingly. The reason you zoom in to focus is that you want to be very sure the subject of your picture looks as sharp as possible. The depth of field is the technical term for the portion of your picture that is in focus. When you are fully zoomed out, the depth of field is large, so many things at different distances away appear to be in focus. However, as you zoom in, the depth of field is reduced so the focusing ring requires fine adjustment. You do not need to know why this happens. It is the effect you are interested in and you will find out how to use it to your advantage when you come to do some clever shots later on. These cameras also have a ring around the lens called the backfocus which you should set periodically. Zoom in on something and focus in the normal way. Then zoom out again and look though the viewfinder to check whether the object is indeed in focus. If it's not, then you have to undo the backfocus screw and adjust the ring a fraction until it is sharp. Do the screw up again to fix it. What you've just done is mechanically set the camera so that objects at a given distance remain in focus no matter where the zoom is. Unless the screw is loose, you shouldn't have to do this very often. But if your pictures are regularly soft then it's either the backfocus or your eyes that need adjusting.

Smaller camcorders do not have a backfocus setting. You basically set the focus by your eye. If it looks in focus then you can safely say it is. No problem.

Zoom

Just a brief word about this because for the time being you will not be using it in a shot. Zoom looks cheap and lazy. What it says is that you the camera operator could not be bothered to get three or four different shots to cut together into a sequence (of which more later). So don't do it. It's not neat and it's not cool. You will be making better quality television than any production that uses the zoom lens in a shot.

However, as well as using the zoom to focus, you will be learning some camera tricks that use the zoom – the fundamental rule about not using zoom in a shot is there to be broken when you know exactly what you are doing.

Shutter speed

This is something we have all seen on television: the shots of computer monitors or televisions with broad lines flickering up or down the screen. In 90 per cent of cases, when you have a television or monitor in shot or indeed if you need to film a screen for some reason, you can adjust the camcorder's shutter speed to get rid of those nasty lines and deliver a rock-steady picture. It can take time to find the correct setting (shoulder-mounted camcorders have hundreds to scroll through) but the result is well worth it and essential in some cases.

What you are doing is no different to adjusting the shutter speed on an ordinary stills camera. Video camcorders take either 25 or 30 still pictures every second (depending on which country you are in) to create the moving picture. The shutter speed is either set at 1/50th or 1/60th of a second. As well as getting lines on computer monitors this means you may also get a blurred picture if you are following a fast-moving object such as a sports car or racing horse. Again, you just have to increase the shutter speed to get a sharper picture but beware that you will also have to open up the iris to let more light in.

Gain

When lighting conditions are low and the iris has been opened as wide as it will go, you have two options to increase the brightness of your picture.

The first is to use more artificial lighting. This can be effective if you have enough lights to balance the light throughout the depth of your picture. However, if you have just one light mounted on the top of your camcorder it can completely flatten the scene and dazzle your subjects giving them reflective faces and pin eyes.

The second option is to electronically brighten the image that's coming into the camera – this is what the gain is for. It's a bit like the night-sight on a gun, only not so deadly. The trade-off from using gain is that the picture quality begins to deteriorate after you have gone about one-third of the way up the scale. By the time you are half-way, it is noticeably grainy. Full gain should only be used for surreptitious night-time filming, like getting pictures of foxes feeding on the heath at night.

Filters

Along with the iris, these help control the amount and type of light coming into the camera. Shoulder-mounted camcorders have a small wheel to adjust for daylight and artificial and fluorescent lighting conditions. They work with the white balance to ensure that the camera records the correct colours. Smaller camcorders probably only have what's called an ND (Neutral Density) filter that simply cuts the amount of light entering the lens. Flick it on when filming on sunny days at the beach or the ski slope. Most cameras with ND filters automatically tell you when it should be turned on.

Batteries

Take a look at the instructions that come with your camera and see what they say about its batteries. Pay particular attention to the rules about charging and discharging. Without fully-charged batteries your filming will not work, so it is vitally important you treat them properly. If the instructions tell you that you should not recharge batteries that are only half exhausted then believe them. The different types of batteries have to be charged in different ways. Ignore this and you may find yourself with a battery that claims it is fully charged but ends up giving you just five minutes of shooting time before it packs up. Batteries are heavy, so you don't want to carry too many of them. But neither do you want to be stranded with no power. Treat them with respect.

Wide-angle lenses

These are not part of the regular kit (and are hideously expensive for shoulder-mounted cameras) but worth mentioning because wide-angle adaptors can be bought relatively cheaply for most camcorders and they should be near the top of your shopping list. Wide-angle lenses make such a difference to your shooting, framing and the general look of your piece because you can fit much more into the picture when getting a scenic shot and you can get much more detail when taking close-ups. Buy one, it'll be worth it.

Zebra stripes

This button is useful for novices. It highlights sections of the picture that are over-exposed or 'hot' with, well, zebra stripes. If you ignore the warning you will find that part of your picture will be bleached white, making the shot unusable. The stripes highlight the fact that there is too much difference between the dark and light parts of the picture. This could include sunlight streaming through a window or a naked bulb indoors or daylight reflecting on a car bonnet. You can either move the camera so the bright light is no longer in the shot or close down the iris so less light gets into the lens. Of course, this second option means that the darker parts of your picture may become too gloomy to see clearly. Zebra stripes are only seen in the viewfinder and are not recorded on tape. When you get to know your camera you will probably want to turn the button off as the stripes can be a bit annoying.

The main points

- Automatic camera functions rarely perform to broadcast standards.
- White balance every time the light source changes.
- Use the automatic iris button to get your setting but keep it on manual.
- Zoom in to focus and then zoom out to frame your shot.
- Never use the zoom within a shot.
- Change the shutter speed to get rid of flickering lines on televisions.
- Use the gain to enhance your picture in low-light conditions.
- Read your battery instructions about charging and discharging.
- Buy a wide-angle lens or adaptor.

3 Shooting skills

One of the most excruciating aspects of the late 1980s was the rise of the home video camera. Suddenly, everybody could afford to take moving pictures of weddings, holidays and family outings when at one time they would have been quite happy capturing the moment with a few well-chosen snaps. Video takes time. Long-suffering friends spend dark hours trapped on sofas because they are too polite to ask for the remote control. Photographs, on the other hand, can be flicked through with interested haste. But why are home videos often so horribly unwatchable? It is not just because they are too long. It is because the camera is rarely kept still. The picture on the screen pans and tracks and tilts and swoops around trying to take in as much of the vista as possible without actually concentrating on anything in particular. The result is that interesting things that are going on in the vicinity keep popping in and out of the frame like rabbits playing in their holes. The eye is never really sure what part of the picture it should be looking at.

So, here is the one golden rule for shooting: do nothing with the camera that your eye wouldn't do naturally.

The shots you should get

Think about what your eyes do when you walk into a room. First, you take in the whole scene. You do not start in one corner and swoop around. Try doing that with your eyes and you will find it is impossible. Your focus jumps from one thing to the next. It is not a smooth motion. So, you take in the whole scene and your eyes don't move. We're going to give that whole scene a name. Let's call it a wide shot.

Second, your attention and gaze is drawn by the things going on in that room. Let's say there is a group of people sitting around the table playing a board game. After taking in the whole scene for a few seconds you will focus on the people round the table. All of them. Let's give this view a name and call it a medium shot.

Then you will move in and look at the detail of what's going on. You will look at the players' faces as they laugh and cringe. You will look at a hand shaking the dice in an extravagant movement. You will look at the dice hitting the table and coming to rest. You will look at the numbers on the dice. You will look at the player reaching for the piece. You will look at the hand moving the dog to Mayfair. You will look at the bank being accidentally knocked to the floor by the player who is losing. You will look at the money and property cards scattering on the floor. These are all individual things that you are looking at and guess what? Your eye jumps between them. It does not swoop. We are going to call them close up shots.

So, when you walk into that room, your eyes take wide shots, medium shots and lots of close ups. When shooting with the camcorder, you will do the same. From this point on, let's do what the entire broadcast industry does and call them WS, MS and CU shots. You can also talk about XWS and XCU shots which stand for eXtreme Wide Shots and eXtreme Close Ups. In our board game example, an XCU would be the numbers on the dice or the bead of sweat trickling down the face of the person who is cheating. The XWS would probably only be accomplished with the aid of a wide-angle lens and a JCB to knock out a spare wall. Clearly there is no XMS because that would be a WS. There, you've learnt your first real pieces of jargon. Go and impress your friends.

The easiest way to approach getting these different shots is to forget for a moment that you have got a camcorder that records its own movement. Instead, think of it as a stills camera and imagine the different photographs you would take to capture all the elements and details of the scene you are shooting. In our example, you would have:

WS	The whole room
MS	Players round the table
CU	Player one scowling
CU	Player two cringing
CU	Player three reaching for the dice
CU	Player three's hand shaking the dice in the air
CU	Dice hitting the table and rolling to a halt
XCU	The numbers on the dice
CU	Player three reaching for the dog
XCU	Player three's hand moving the dog to Mayfair
CU	Player one knocking the bank
CU	Money and cards scattering on the floor
XCU	Bead of sweat trickling down player one's forehead

What you will notice about all of these shots is that, even though we are viewing them as stills, there is inherent movement to them. The camcorder is being held still to keep the point of view in the same place for the duration of the shot. It is the subject that we want to see movement from. You do not want to shoot a series of static images where nothing happens because it will be boring to watch and the viewer may as well look at the pictures in a Sunday supplement instead. You actually want as much interesting activity and movement as possible going on in the picture. This activity is what the viewers will focus on when they are watching the television screen and they will want to see the detail which is why we would get all those CU shots.

Within the context of our game, we could also get shots of the clock on the wall (MS of the whole clock and CUs of the pendulum swinging, the second hand moving and the cogs turning). Maybe there would be a coffee machine brewing a pot of Columbian (MS of the machine and CUs of the water dribbling through and steam escaping from the top). Or a spectator dozing (MS of the person and XCU shots of eyes and ears twitching). But rather less interesting would be shots of books on shelves because they just sit there. Or a pile of magazines for the same reason. Or a telephone on a table. We want to see activity within the still picture.

There is one major exception to the not-moving-the-camera rule. The tracking shot. If the subject of your shot is moving, a driving car or walking person for instance, then you can follow them with the camera as long as they remain in the same place within the viewfinder for the duration of the shot. When you see the picture on screen, your eyes will remain in the same place on the subject. In effect, the camera is not moving. The background is. It is not easy to take a slick tracking shot. For a start, you have to hold the camera by hand meaning there will be some shake. Second, you have to move at the same time and speed as the subject to make sure they don't end up darting all over the frame. Good tracking shots can be very effective. And sometimes, for instance if you are filming a defendant arriving at court, you may have no choice.

The shots you shouldn't get

If you have ever watched any television in your life, you will know that there are other shots. Pans, for instance. A pan is usually a WS that starts at one point in the scene and moves round to another point in a single flowing action. This is something your eyes cannot

The smooth tracking shot

What is it?

Tracking shots follow moving objects. The idea is that you keep the object in the same place within the frame while the background rushes past. Peter Greenaway used them a lot in *The Cook, The Thief, His Wife and Her Lover* when the waiters were taking the food from the kitchen to the dining room to the tunes of Michael Nyman.

Unfortunately, hand-held tracking shots always wobble and jar. Much as your arms try to absorb shocks, every step you take is another bump to the camera. Traditionally, you could only do the tracking shot by laying railway tracks for the camera to glide on or by hiring a Steadycam operator. However, there is an alternative for paupers.

How do I do it?

As you have probably noticed, office seats and supermarket trolleys have wheels. To do your tracking shot you sit on the chair, locking the camcorder in your lap, while a helpful assistant pushes you along, following the action. Or you rest the camcorder on the trolley and push it yourself. Either way, your tracking shot will be as smooth as a timeshare sales clerk and everyone will wonder how you did it.

do. Much better to get an WXS of the whole scene to establish where everything is in relation to each other and then MS and CU shots to pick out the detail. If, for some reason, you feel you absolutely cannot do without the pan (okay, sometimes you will not be able to get a sufficiently XWS that shows where everything is) then make sure the beginning and end of it concentrate on interesting things and hold these shots at either end of the move for at least ten seconds.

Similarly, the tilt. Either up or down. This is usually used to show how tall a building is. Starting with a MS of the entrance and ending with an XWS of the front of the building from the same point on the ground. There is nothing wrong with either shot in its own right. It's the movement in between that grates because, again, your eyes wouldn't do it.

Finally, there is the crash zoom. We have already established that using the zoom in a shot is a bad thing. Crash zooms are when the shot starts with a CU of some interesting detail and then, in an instant, expands right back to reveal, hey, it's a Christmas tree we are looking at. Unless you are working on some particularly esoteric television station that assumes its audience's brains already view the world like that, then don't use it.

For all these shots, a bit of time, effort and thought will provide far better alternatives.

Using a tripod

Holding a camcorder rock-steady for the duration of the shot is an act that even those who have given up coffee cannot achieve, particularly as camcorders are getting lighter. Thankfully, there is an easy solution – the tripod.

They may be a pain to cart around and they may take a few minutes to set up. But they are vital for producing professional pictures rather than the sort of home movies Uncle Billie shoots. The tripod is very important. Your pictures will be rock-steady.

Tripods come in a lot of shapes and sizes at many different prices. There is a balance between selecting a tripod that is too heavy for you to carry around on your own (with all your other equipment) and one that will not adequately support the camera. You want the lightest one that will effectively hold the camera's weight. Treat your tripod with love and care. You do not want to overtighten the bolts because the threads will eventually go. Nor do you want to yank the head to which the camcorder is attached because it will eventually wear and wobble. Camcorders are secured to tripods with plates that screw into a hole on their underside. This hole is a standard size. The plate then clicks into place on the head of the tripod. It is easy to lose the plate, so don't.

When the camcorder is clicked into place, you have to make sure it is level. Often you can do this by looking through the viewfinder and seeing if the horizon or some other horizontal point of reference simply looks right. More expensive tripods have a small spirit level built into the head. You adjust all the legs until the bubble sits in the centre.

An alternative to the tripod is the monopod. Instead of three legs it has just one. Clearly, it cannot stand alone. It is intended to give you a bit more stability than holding the camera by hand but will never give you the really steady shots you will get from a tripod. Monopods are smaller and lighter than tripods to carry around but should only be used if a tripod is totally out of the question.

Composing your shots

Now you know what kind of shots you should be getting, you need to start thinking about how you are going to compose them. You have to work out what elements of the scene you are going to include in the shot and where are you going to put the camera. This is called framing.

If you take good photographs then you should have no problems taking good television pictures. Composing shots is largely an instinctive process. Some people take good holiday snaps. Some people don't. It is the same on video. Clearly, every shot you take is different. But there are some common points you should remember.

Balance the light

Video camcorders are not very forgiving when it comes to dealing with extremes. Very high lighting conditions need filters. Very low lighting conditions need gain. Imagine the problems you will have if there are both bright lights and dark shadows within the same picture. Windows cause this very problem during the day. Your eye may be able to look out of a window and see exactly what is going on without the room you are in turning black. But camcorders cannot do this. If the iris is set for interior lighting conditions, then any light that strays in through a window will overexpose the part of the picture it hits. If there is direct sunlight streaming through, the entire area will be bleached. Similarly, if you are outside during the day and the iris is set to cope with sunlight, you should expect to see the shadows under trees turn dark.

What you want to do is ensure that the lighting conditions throughout the picture are as evenly balanced as possible. Do not frame your exterior shot with half of the picture in the light and half of it in the shade. Position yourself and the camera under the tree, for instance, and take a shot that is filled with daylight. Do not include a window in your interior shots if there is sunlight streaming through it. Instead, position yourself with your back to the window and fill your shot with the darker room. Or you could close the blind. For the same reason, you should not shoot an interviewee either in front of their office windows or with the sun hitting parts of their face.

Give pictures depth

Brick walls are not that interesting. Neither are office dividers. But busy streets are. And airport check-in areas are. When framing shots, your aim is to make it as visually interesting as possible. And that means thinking about the background. You may have the most interesting and exciting subject in the foreground but the pictures will still end up looking boring if you are shooting against a grey wall. However, if you are shooting in the middle of a milieu and the background is made up of things happening way into the distance then your shot will end up being a lot more dynamic.

Imagine you are shooting a racing car having its tyre changed by the side of the track. From one angle, you will get the side of the pit stop as your background. Boring. From another angle you will get cars racing round the track. Interesting. Similarly, a shot of a street sign will be very lifeless if taken from straight in front. Place yourself so that the right-hand side of the sign fills the right-hand side of the frame and point your camera down the street. The sign will extend about halfway across the picture and the activity in the street will fill the rest. Figures 3.1 and 3.2 show how this works.

You do not have to worry about the background taking precedence. Your foreground subject will remain the object of the viewer's attention not least because the stuff going on in the background will be slightly out of focus. But the depth it provides will make your shot look better.

Think about perspective

The most boring shot in the world is a MS taken at shoulder height as demonstrated in Figure 3.3. It sits there on screen like a pudding failing to rise. With a little thought and ingenuity, you can take pictures that not only depict the action but look interesting as well. Now you do not want to go overboard and have every shot looking like it is an out-take from a David Lynch movie. But a little variety never did anyone any harm. A wide-angle lens helps immeasurably when you are framing unusual shots. Here are some examples.

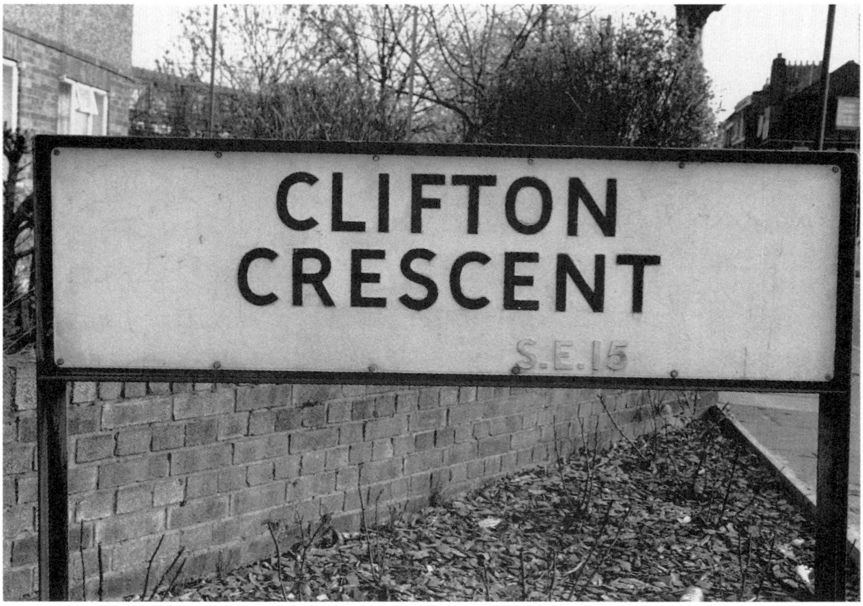

Figure 3.1
Well, we get the picture . . .

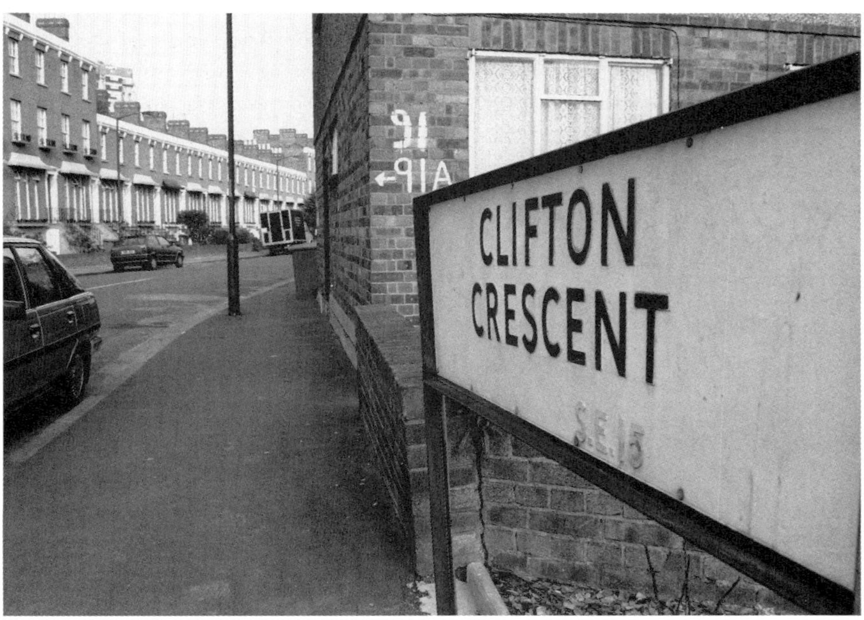

Figure 3.2
. . . but isn't this a bit more interesting?

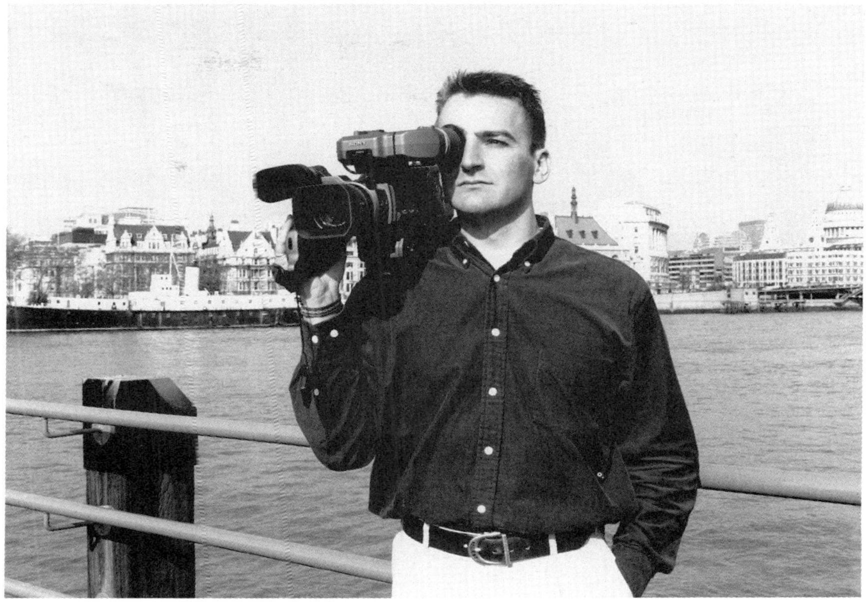

Figure 3.3
It's an obvious way to hold the camera but produces boring shots

Cars driving along the road

Shot at shoulder-height, they simply drive past. However, crouch right down as near to the edge of the road as you dare and frame the shot so that the width of the road fills the width of your picture. You will see the other end of the road trailing off into one corner. When cars approach they will start as little dots in the distance. Gradually, they will move closer and closer until they fill the entire frame as they roar past. It looks fab. You can do the same with people walking along a path but they move a lot slower. Figures 3.4 and 3.5 show how to get this shot right.

A room

As we have already identified, in any room it is generally difficult to get a good WS because you will probably need to remove a wall. However, find some stepladders or a table to stand on and get the camera as high as you can in one corner of the room. Take a look at Figures 3.6 and 3.7. You will be amazed what a difference this bird's-eye view of the scene can make.

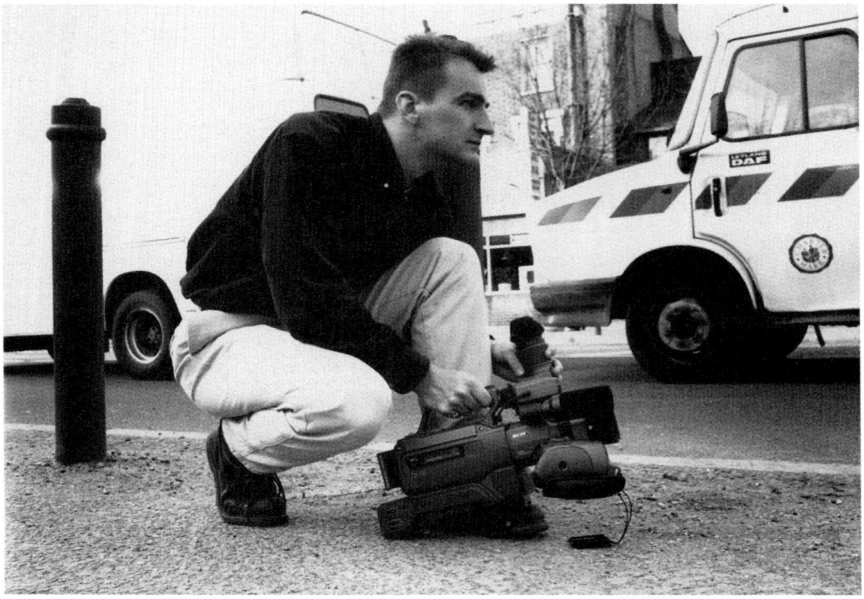

Figure 3.4
Holding the camera close to the ground and angling it up . . .

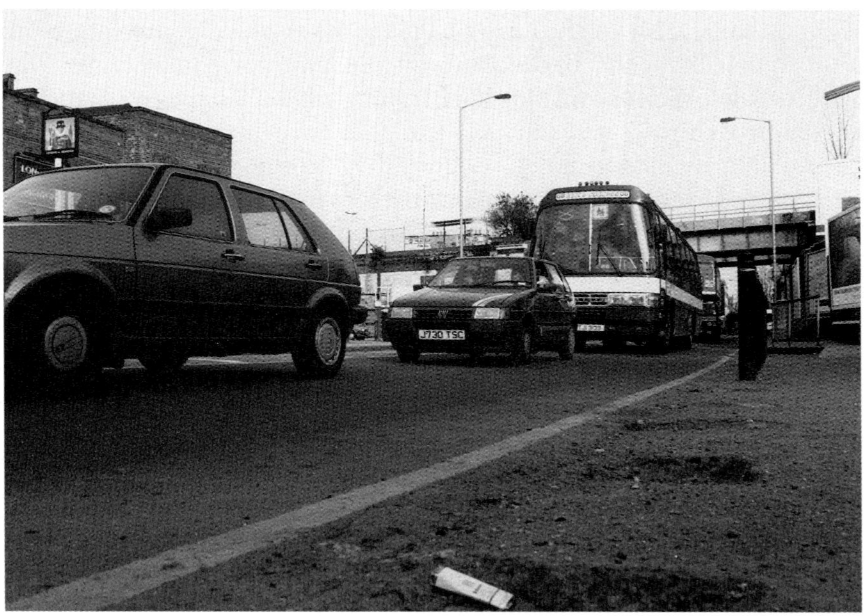

Figure 3.5
. . . will mean your WS of the road will be beautifully framed

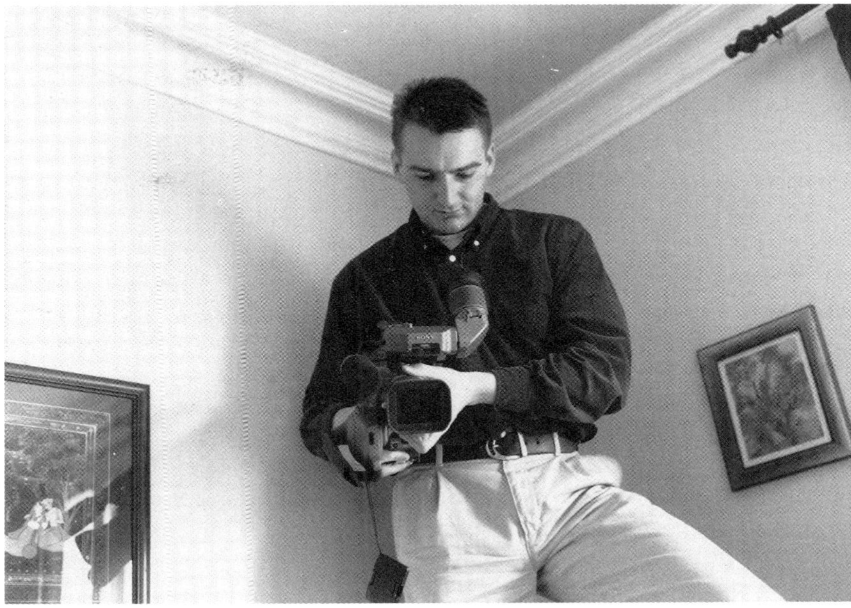

Figure 3.6
Finding something high to stand on . . .

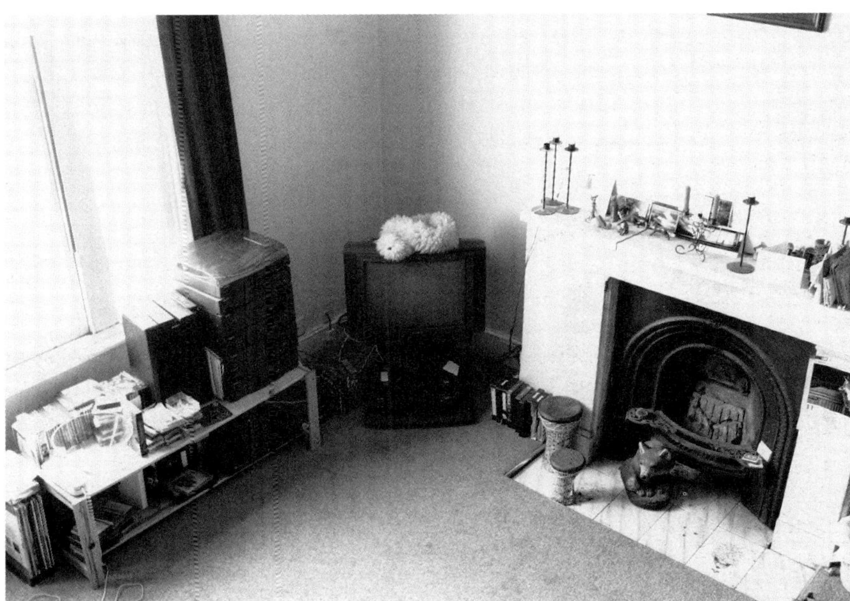

Figure 3.7
. . . will make your WS of the room appear a lot fuller

A person playing cards

So, you've got the CU of their wrinkled brow and the MS of them sitting there but that's about it. How about positioning the camera underneath the table and pointing it up between their legs so initially we see the cards in the bottom half of the picture and an up-the-nose shot of the face. The player then selects a card and flicks it to the table right across the frame. Hey, you could even slow that bit down in the edit to make it even more cool. It doesn't take too great a leap to imagine how this kind of shot can be applied to anything from operating computers to washing up.

Reflective surfaces

These always good for a weird shot. The backs of spoons may be too rough but you can bet the chrome on a car bumper would show a nice reflection of someone walking away if you can find one. I once took a great reflection shot of a chef beating eggs from the metal surface in a hotel kitchen. A slight word of warning though. Mirrors are a bit clichéd now for reflection shots.

Fridges, cupboards and kilns

There is always one more shot to be taken after you shoot a closed door being opened. The reverse one looking out. So, how do you do it? Start the camera rolling, position it in the back of the fridge pointing out and close the door. Your subject opens the door, retrieves the cream cake and closes it again. You've got your shot.

Fill the frame

You do not want to see large patches of nothing on the television screen so you have to beware of concentrating the action in one corner or side of the frame. Watch out for:

Sky

It's interesting if the sky is deep blue and there are white clouds scurrying across it, but boring if it's uniform grey.

Roads

Roads are interesting at the right angle with colourful cars roaring past, but boring if they are empty.

Walls

There isn't a lot you can do with walls. It's best to move in closer to the subject or turn around and get another angle.

Surfaces

Again, not a lot you can do except reframe the shot with the camera in a slightly different place.

As a general guide, you don't want more than about a quarter of the frame filled with nothing. Reposition the camera if you find that it is.

Hold your shots

The final point to make about shooting is that even though you are treating the camcorder as a stills camera, you should hold the shots in position for at least 20 seconds, preferably more. Why? Continue reading and you will find out.

The main points

- Do nothing with the camera that your eye wouldn't naturally do.
- Treat the camcorder as a stills camera and do not move it.
- Hold your shots for at least 20 seconds.
- Keep the subject at the same point in the frame when getting tracking shots.
- Use a tripod as much as possible.
- Balance the light within the frame.
- Give your pictures depth with an interesting background.
- Get interesting shots from unusual perspectives and fill the frame.
- Avoid panning, tilting and crash zooming.

4 Directing

Imagine the credits of a feature film. Who's the top dog? That's right: the director. So it must be a highly skilled job. But look further. There is a set designer, costume designer, director of photography, make-up artist. You get the picture. In a feature film there are countless other important jobs that contribute just as much to the look of the finished product as that of the director.

For the features and documentaries you will produce as a videojournalist, you do not need all those other people. You are filming real people doing real things in real locations. The action is already happening. They don't need different clothes to those they are wearing, nor do they need special make-up. Your job is to capture everything on tape and turn it into television.

In videojournalism the distinction between the traditional jobs of shooting and directing becomes blurred into one process. Indeed, you can work quicker than a conventional director and crew for this type of television because you do not have to communicate to anyone else what it is that you want. You just do it yourself. This is especially important when events are unfolding before your eyes and you have to work pretty darn quickly to make sure you have everything you need in the can.

Having established that you have the technical know-how to shoot broadcast-quality pictures (they are well-framed, in focus, the white balance is correct . . . that sort of thing) you now need to start looking at how you are going to use those pictures to tell the story. And there is no essential difference between you doing this and a movie director doing what they do. They just have a lot more money and resources at their disposal.

This is directing

Here is what I mean. Let us take a very simple example. You are doing a story about an athlete training for some contest and one of the locations in which you have to shoot is the gym. The athlete is pumping away on the step machine. You get these shots:

XWS	The gym
WS	Athlete on step machine
MS	Athlete from waist up
CU	Sweaty face bobbing up and down
CU	Steps pumping up and down
XCU	Hand gripping the foam bar
XCU	Step machine electronic display

The athlete is probably on the machine for a while meaning you with your single camera can get all of these shots quite happily in the time provided. The final shot of the electronic display will probably have to be taken after the athlete has finished. You simply set the camera up on the tripod and pump the steps by hand to make the display move.

You now have to decide what order you want these shots to be seen in. Here is a piece of free advice: start with your most interesting shot. There are a lot of channels out there. A lot of other competing programmes. This is not a movie in a cinema. The viewer has not made a personal commitment to sit down and watch your work. You have to grab them as soon as possible. And that means right at the top.

You also want people to know where you are and what is happening quite early on so the establishing XWS needs to go near the top too. How about this order:

XCU	Step machine electronic display
MS	Athlete from waist up
XWS	The gym
CU	Sweaty face bobbing up and down
CU	Steps pumping up and down
XCU	Hand gripping the foam bar
WS	Athlete on step machine

That would work. What you have done is to create a sequence. And that is what directing is all about. You only have one camera so you can only record one thing at a time. Your job as a director is to make a plausible continuous flowing sequence out of a collection of shots that were taken at different times. Sometimes, the shots were taken at wildly different times. But you have to put them together in some sort of order so that they work as television.

Building tension

Let us go back for a moment to the people playing the board game. We have already established a number of shots can be taken of a number of different things. The game itself provides us with the most scope. But there is also the clock, the coffee machine and the sleeping spectator. Altogether we have about 20 individual pictures to play around with. We need to put them into a sensible order that tells some sort of story. How about this:

CU	Water dripping through the coffee machine
MS	The coffee machine steaming
WS	The room establishing the people and the coffee machine
CU	Bead of sweat trickling down player one's forehead
MS	Players round the table
CU	Player three reaching for the dice
CU	Player two cringing
CU	A hand shaking the dice
MS	The clock on the wall
CU	The pendulum swinging
CU	The second hand reaches the top of the hour
CU	The dice crash to the table
XCU	The numbers on the dice
CU	Player one scowling
CU	A hand moving the dog to Mayfair
MS	Spectator asleep
XCU	Spectator's ear twitching
CU	Player one knocking the bank to the floor
CU	Money and cards scattering on the floor

You can see the story being told. The coffee, the bead of sweat and the clock all help build tension. Clearly, it's a key moment in the game and one of the players doesn't like how it's going. But not everyone there thinks it's so exciting.

As before, we started with an interesting shot and soon came to the establishing WS. It would be no use going straight from the CU to the WS because it would probably not be clear where the CU is in relation to the WS. If you think this example is unlikely for news or documentary production, what if your story is about a board games championship? They do happen. And the order your pictures have been seen in has said far more about the tension in the room than any number of words in a script.

Playing with time

Creating sequences is also about playing with time – invariably compacting it. Let's say you were shooting someone making a cup of instant coffee. It takes about 20 seconds to fill a kettle from a tap and minutes for the kettle to boil. You do not want your piece to take so long. This is more like it:

CU	Tap being turned
MS	Kettle being filled
WS	Person filling the kettle
CU	Kettle sliding into its holder
XCU	Switch being pressed
CU	Mug being placed on surface
MS	Instant coffee being put in mug
CU	Kettle boiling
MS	Boiling water pouring into mug
CU	Person taking a sip

Take a look at Figures 4.1 to 4.10. They show precisely this sequence of shots. You can imagine how they would run together to make about 25 seconds of flowing pictures on screen. If only making coffee were so quick in real life.

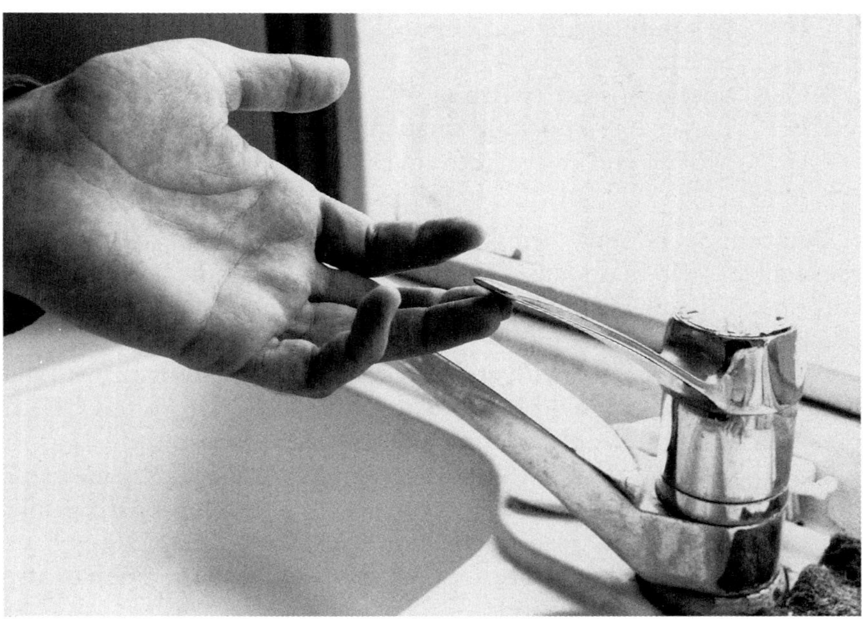

Figure 4.1 CU Tap being turned

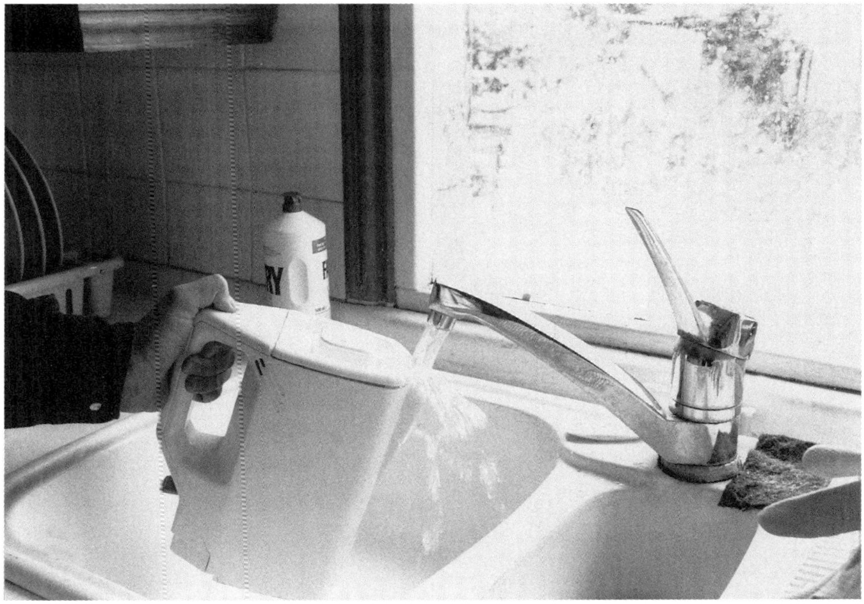

Figure 4.2
MS Kettle being filled

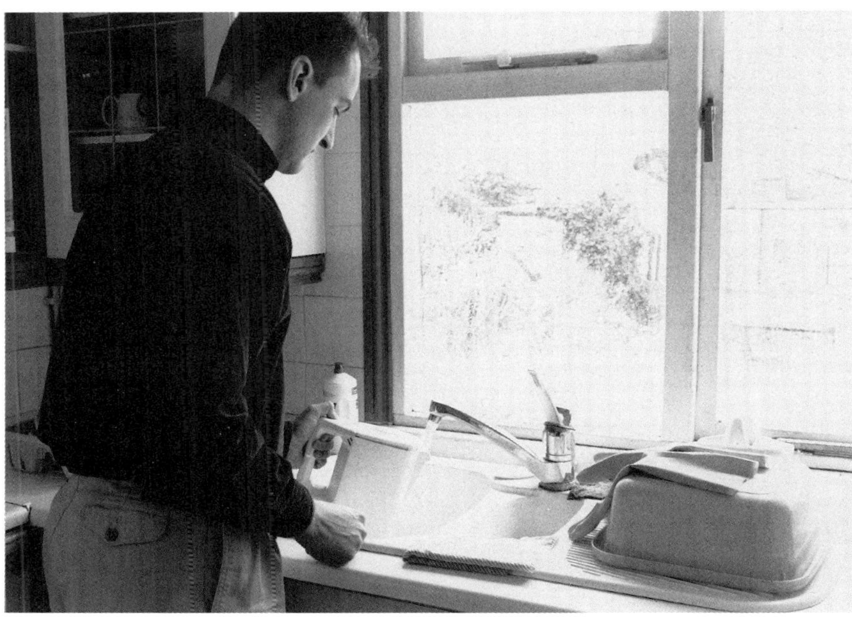

Figure 4.3
WS Person filling kettle

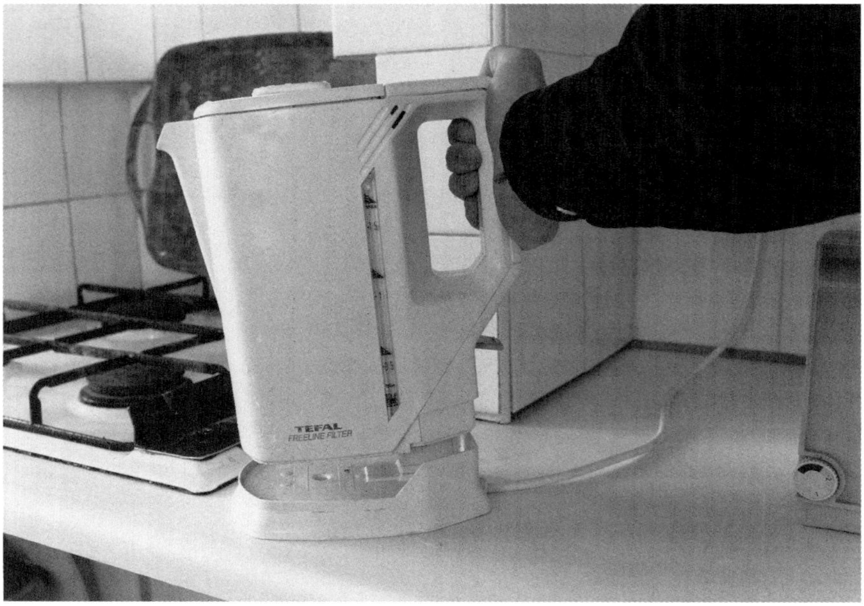

Figure 4.4
CU Kettle sliding into its holder

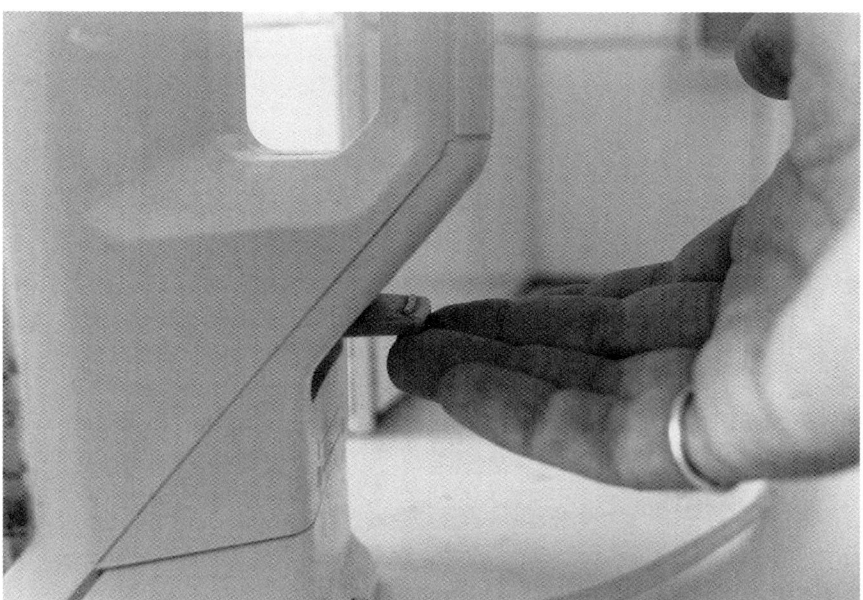

Figure 4.5
XCU Switch being pressed

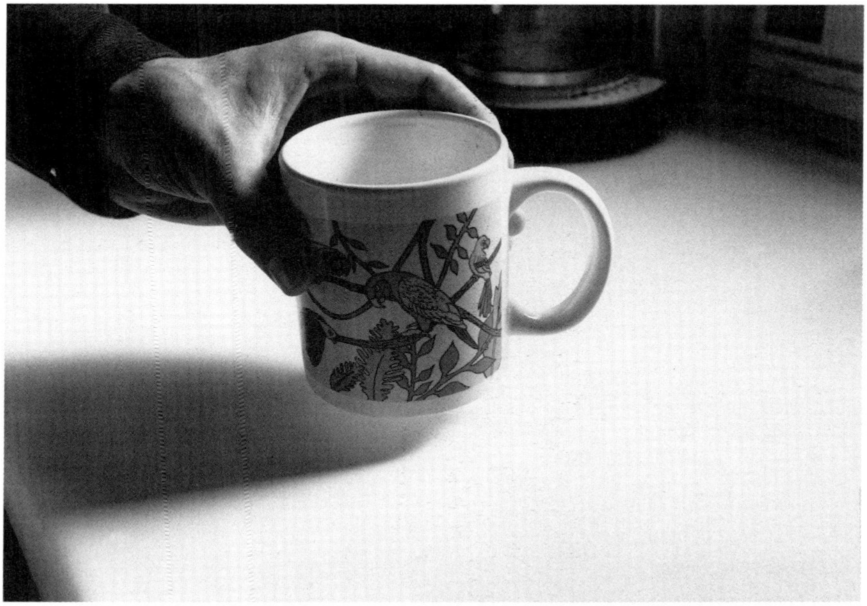

Figure 4.6
CU Mug being placed on surface

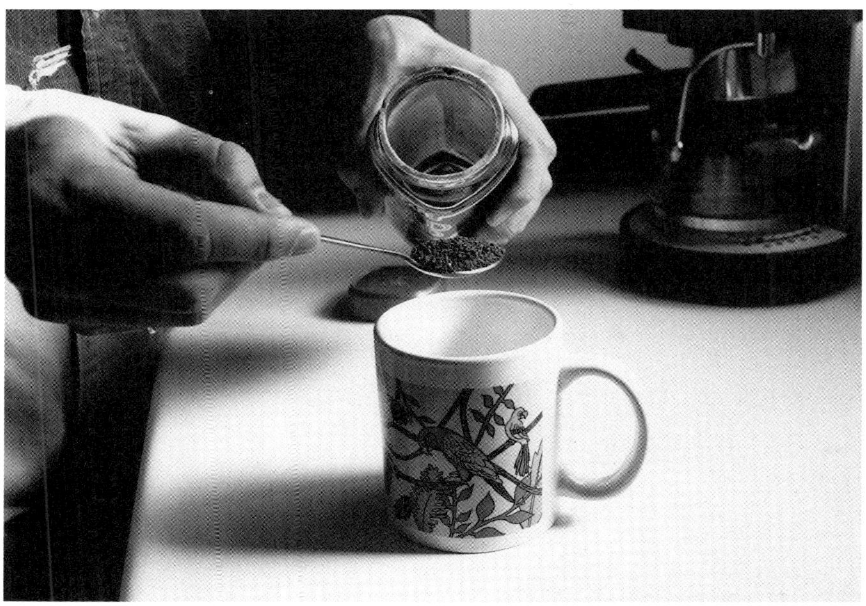

Figure 4.7
MS Instant coffee being put in mug

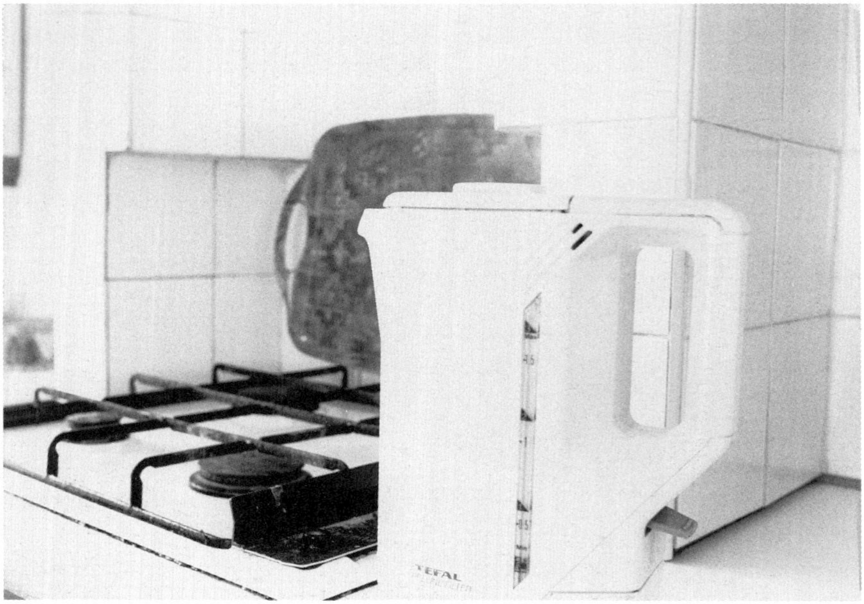

Figure 4.8
CU Kettle boiling

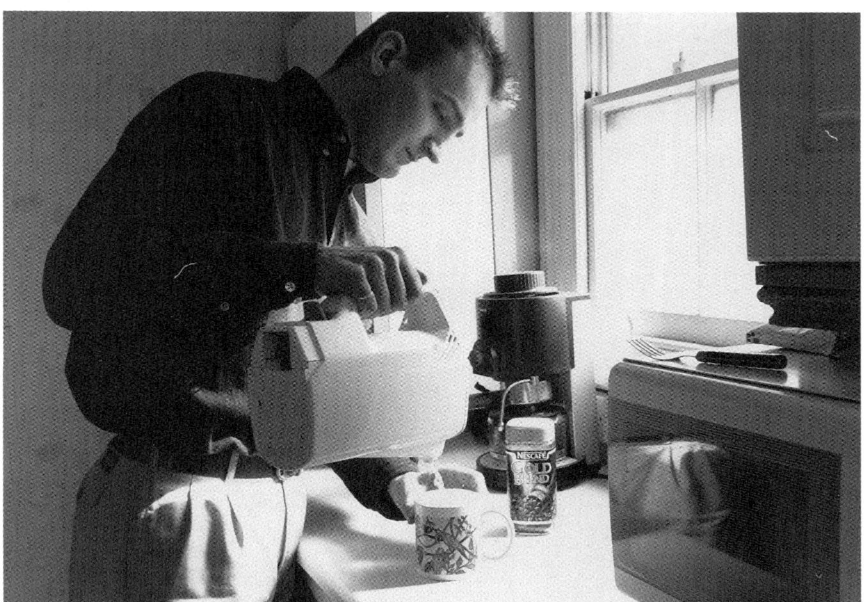

Figure 4.9
MS Boiling water being poured into mug

Figure 4.10
CU Person taking a sip

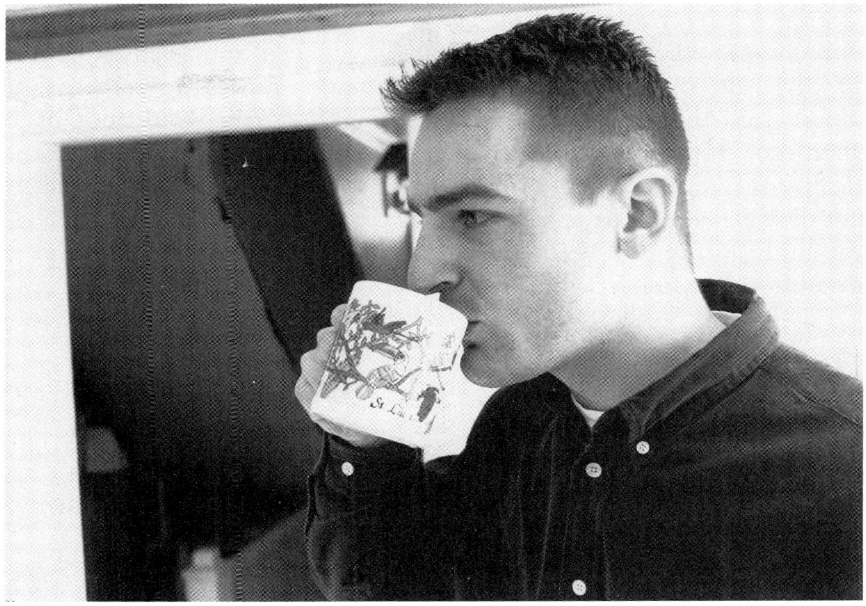

Transitions between shots

The transition between one shot and the next is the most important aspect of your quest to create the illusion of continuity. It's called a cut. Each cut has to look like it actually occurred in real time. In effect, it should look like you had two cameras both filming the same thing from different angles. You know you didn't. The people you are filming know you didn't. But the viewer is unaware.

So when we go from a MS of the kettle being filled to the WS of the room, the water has to remain flowing and the hand holding the kettle has to be the same. The cut wouldn't work if the water was flowing in the MS and not in the following WS. The water in this case is an important point of reference from one shot to the next and it has to be doing the same thing both before and after the cut.

Going back to our athlete in the gym. The feet, knees, hands, body and head are all important points of reference in the sequence. And they have to be in the same place between adjacent shots. The athlete's body will be leaning on one foot in the MS from the waist up so it has to be leaning the same way in the following XWS. Similarly, the body has to be in the right place for the next cut to the CU of the sweaty face bobbing up and down and the next CU of the pumping steps.

This is why you hold your shots for at least 20 seconds. It gives you choices when making your cuts. If you had only held each shot for the four or five seconds that ultimately appear on screen you would find that they would not cut together. You would not have shot the important points of reference in a sufficient number of positions within the frame. You would find yourself being forced to cut between a MS of the body leaning on the left foot and an XWS of the body leaning on the right foot. This is called a jump cut and it doesn't work.

Moving objects in and out of the frame

You can also make this job easier for yourself by removing the points of reference from one cut to another.

Take the transition between the WS of the kitchen and the CU of the kettle being slid into its holder. We have compacted the time it takes to turn off the tap and walk over to the kettle holder. The WS will show where the holder is in relation to the tap. But if the CU of the kettle holder does not have the tap or the person in the background then there are no points of reference between the two shots. So as long as there is a moment's hesitation between the WS ending and the kettle sliding into its holder then the cut will work even though in reality it probably took ten or so seconds. The reason it works is that the kettle is moving into the frame and who knows what happened to it before it got there.

So when filming people or objects that are moving always try to let them move into or out of the frame even if this means holding the shot for longer. If they move into the frame and out of the frame in the same shot then you have the ultimate flexibility to use the start or the end of it.

Take a standard shot of someone walking out of the frame. The next shot of them can be in a completely different location. It will work. But if you did not let them move out of the frame and the next shot was a completely different location then everything around them would jump and the cut wouldn't work. Unless, of course, you cut to them walking into the frame.

Cutting away from the main action

If you are doing a story about a car crash and a rescue truck has arrived to take the vehicle away there will probably be a group of morbid spectators gathered around to view the spectacle. You may find you miss some of the shots you strictly need to create a flowing sequence out of the truck lifting the car to its trailer. So, you make sure you get plenty of shots (WS, MS and CU) of the watching crowd.

The car and truck are nowhere to be seen in these pictures. Anything could be happening to them while we are not looking at them. So you can cut quite happily from one picture of a rope being attached to the car to a shot of the on-lookers gasping to a shot of the car safely on the back of the truck.

The shot of the onlookers is known as a cutaway because it cuts away from the main action to something that does not feature its reference points. When we return to it, the main action can have changed drastically and it will work.

Cheating

Black cabs, London buses and firefighters in uniforms all look pretty similar to each other unless you get down to the real detail of the number plates or the faces. So you will probably find that you can happily cut between shots featuring different subjects as long as the main reference points are the same. Buses and taxis are the same shape and colour. Firefighters wear the same uniforms and helmets. You could have five shots:

WS Back of the firefighters
MS One firefighter struggling with the hose
CU Feet attempting to get a grip on the ground
MS Firefighter crouching down with the hose
CU Hands gripping the end of the hose

Not one of these shots uniquely identifies an individual firefighter so they could all be different people and still the sequence would cut together and appear continuous.

The line and how to avoid crossing it

One final important point about directing is the much-mentioned line. This is something you should not cross. But what is it?

When you glance around yourself, you are looking at objects and actions from one perspective. That perspective does not move unless

A sparkling landscape

What is it?

A whole night sky's worth of shining stars atwinkling on the surface of the lake. Eight points of light spearing across the screen from a central crown in the forest trees. You get the picture. But do you really have to employ a seventies mirror globe or carry a complicated array of light-twisting filters to get that sparkling effect? Of course not.

How do I do it?

Get your camera and point it directly into the sun. Close the iris right down and open it just a fraction. Some cameras that use electronics to control the iris will not be able to cope. But if you have a good camera, you will see the sun turn into a star that would rival the one over Bethlehem. Alternatively, position yourself on the side of the lake and point the camera at the surface so the almost-closed iris catches the glints of the sun reflecting off the ripples. With swans and ducks milling around, you will get beautiful sparkly pictures.

you do. However, the camera does move. It gets different perspectives. And on its own, it doesn't let the viewer know how that perspective is changing. If you take a series of shots from every conceivable angle and cut between them indiscriminately, it would be very easy to disorientate viewers so they lose their sense of direction. To avoid this, you have to make sure you don't cross the line.

The line is an axis and every scene has one. A few examples:

A computer operator – the line runs straight through the computer following the person's line of sight.

Two people talking – imagine them directly facing each other. The line runs through them both. Figures 4.11 through 4.14 show why you have to remain on one side of their conversation.

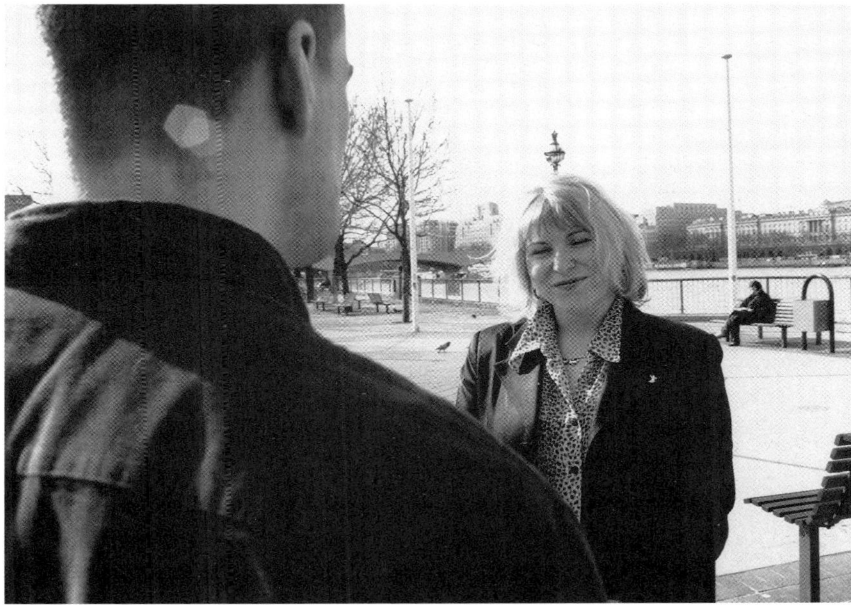

Figure 4.11
Over the inverviewer's shoulder

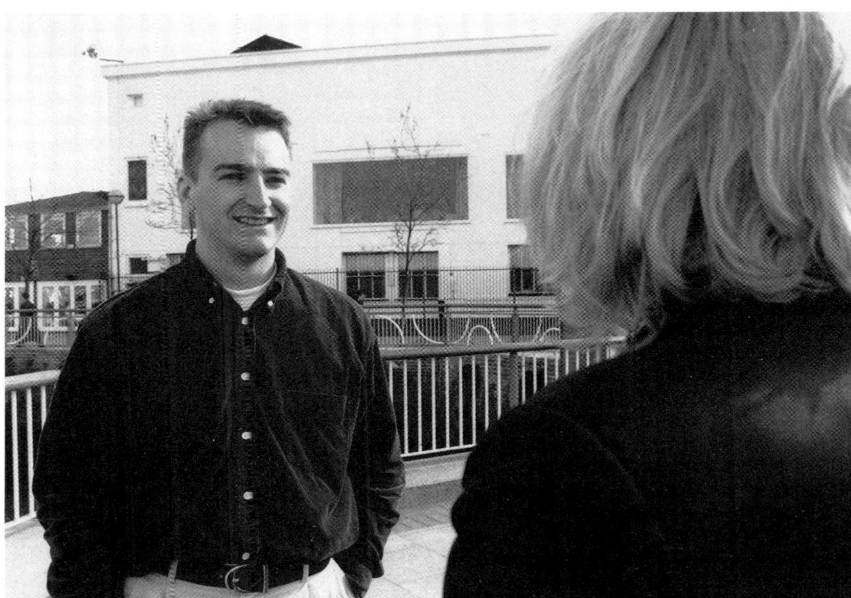

Figure 4.12
Over the interviewee's shoulder

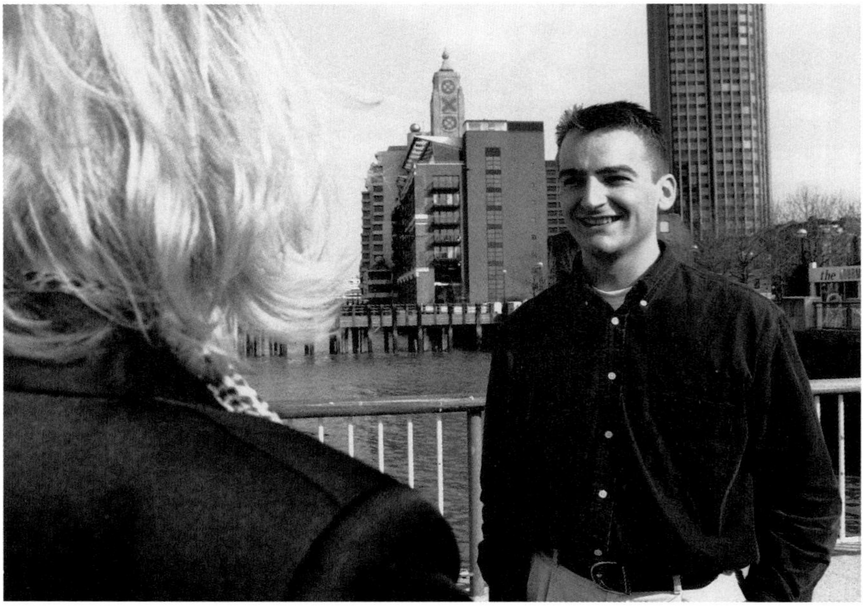

Figure 4.13
It looks like they've swapped places

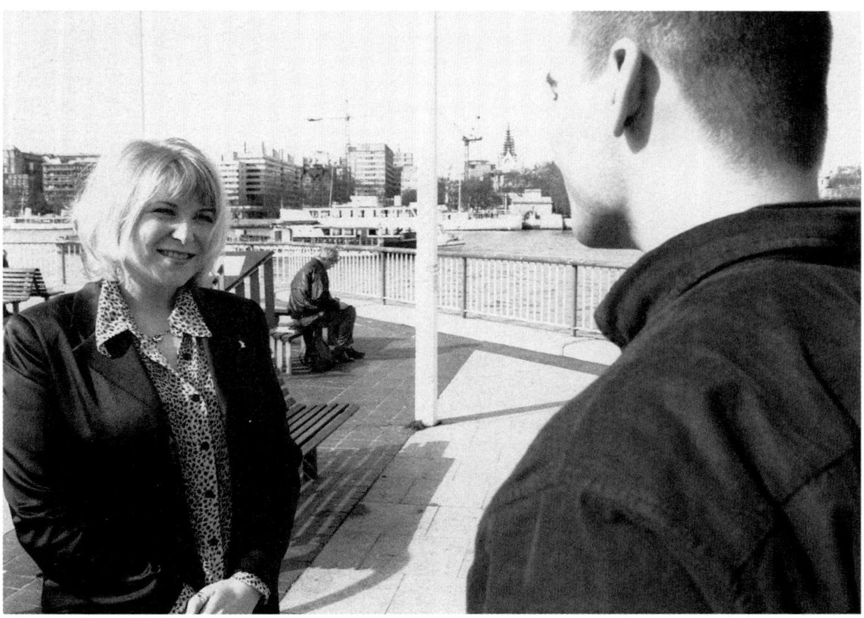

Figure 4.14
But all you've done is crossed the line

Firefighters putting out a blaze – the line follows the firefighters' line of sight through the burning house.

A person walking down a path – the line follows the direction they are walking in.

You have a 180-degree arc on your side of the line so there is plenty of scope for taking pictures of the action without needing to cross it. If you do inadvertently cross the line, it will look like the objects or people have shifted position. Someone who is walking from left to right will suddenly be walking from right to left. An interviewer and their interviewee will appear to have changed seats. The firefighters will be putting out a blaze on the opposite side of the road.

Imagine a round table featuring eight card players. Where does the line run? A complicated one. There are potentially eight different 180-degree arcs to choose from. The answer is: the line depends on the previous shot.

In this example, you can change the line mid-scene by taking a head-on shot of one of the card players and cutting to a second shot of their point of view. Your new line can now be on either side of that picture.

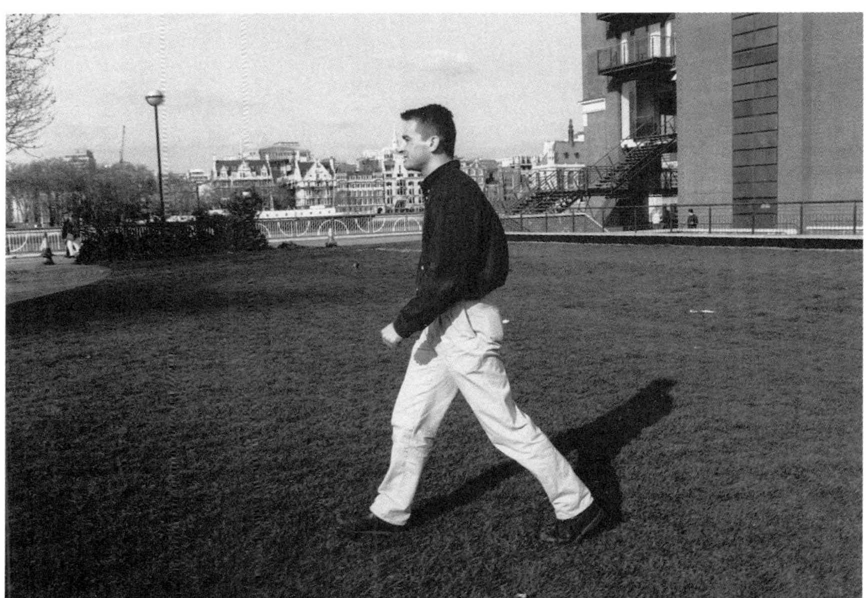

Figure 4.15 The first shot determines where the line is

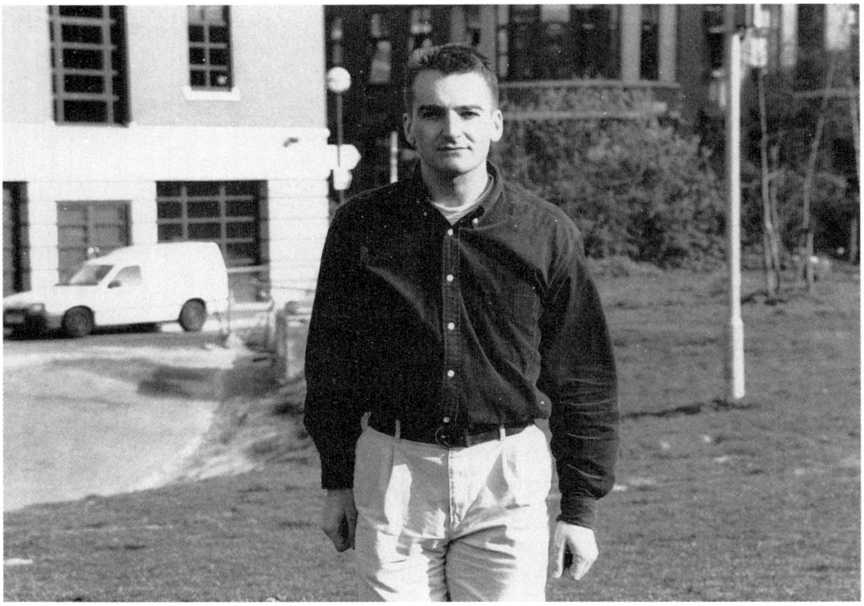

Figure 4.16
The second shot is taken directly on the line

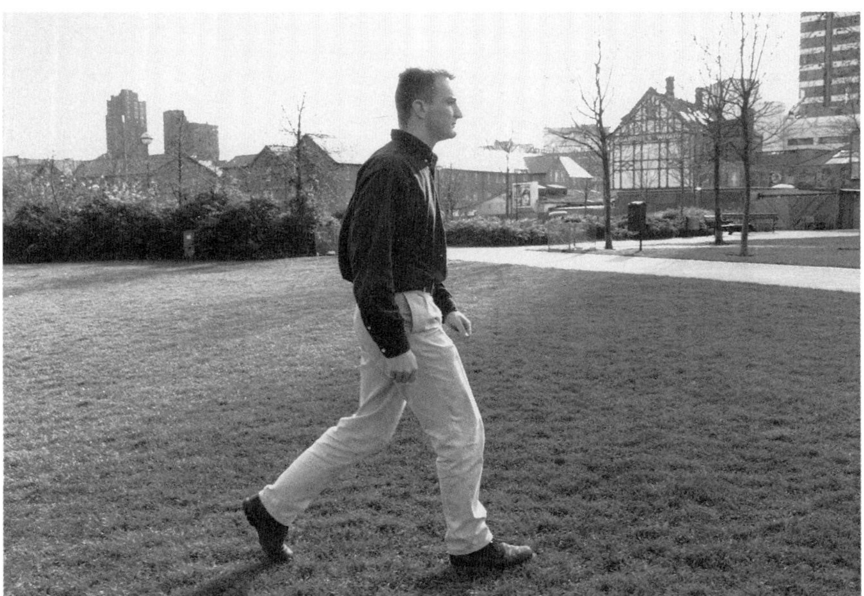

Figure 4.17
So you can move across it for the third shot

Take your person walking down a path. You can cross the line if you also take a head-on shot of them walking directly towards you. Or away from you. Like this:

WS Person walking right to left
MS Person walking directly towards camera
WS Person walking left to right

These shots are shown in Figures 4.15, 4.16 and 4.17. If you run them in this order, you'll make it quite clear how the perspective is shifting when you cross the line in the scene.

The best way to learn exactly what the line involves is to deliberately cross it in a series of pictures. You will see how confusing it is if the perspective leaps all over the place when you try to cut those pictures together.

The main points

- Think about ordering your shots while you are shooting them.
- Put your most interesting shot right at the top of the piece.
- Put an establishing WS near the top of the piece.
- Let objects move into and out of the frame.
- Remember to shoot cutaways from the main action.
- Work out where the line is and do not cross it.

5 Shooting a package

There is a wonderful tale about a senior manager at a television station who had come up through the ranks in the newspaper industry, achieving his superior broadcast position without knowing a frame about making television. Early on, someone mentioned that an important news story was coming in. The package would be there by 15.00. In good time, the manager sent a minion down to the reception desk to make sure that the moment the brown paper package was delivered, it could be brought immediately to the newsroom for transmission. What fun they had with that one. Because a package in television terms is nothing to do with delivery bikes and padded bags. It refers to a completed story that is ready to be broadcast. The senior manager has now moved stations but it is still not known whether he knows anything about making television.

Packages contain a number of basic elements: pictures, sound, script, voice-over, quotes, pieces-to-camera and maybe even graphics. When the news presenter refers to the roving reporter and then looks to the imaginary television placed just to the left of the camera, they are introducing a package. Packages are what you will make.

Now you know all there is to know about shooting and directing, you can be sent out to get the rushes for a package. You see, you are starting to get to the meat of making television so there are lots of technical terms to learn. Rushes refers to the unedited video tapes you bring back from your shoot. Tapes chockfull of shots lasting more than 20 seconds. Those are rushes. And the first thing you are going to do is get the rushes for a simple one-minute package.

What you should choose

Let us for a moment forget the fact that you will have to write a script for this package and instead concentrate on the pictures. This is the most important aspect of doing stories on television, and one that print or radio journalists have to get their minds around. In television, if you don't have the pictures, you don't have a story.

The more skilled you get at making television, the more inventive you become at finding good pictures for stories that would ordinarily have little or no visual value. But at the moment, you simply want to improve your directing and shooting skills. You also want to make things relatively easy for yourself. So you should choose something that is inherently televisual. Look at what we have already examined: a game being played, an athlete working out in the gym, someone making a cup of coffee, firefighters battling a blaze. Now I don't suggest you set fire to a house and set yourself up waiting outside, but let's expand a bit and think of other subjects which are easily accessible that you could select as your first real exercise in shooting.

You want to choose something where people are doing things, something which involves action and something that will allow you to get a lot of shots of many different things in one location – because we do not want to make a meal out of this.

It is probably easiest if you choose a single subject. A person who you can profile. So that person should be doing something that has many stages to it. So, what about: someone in a pottery class, a chef in a restaurant, a cobbler mending shoes, a gardener, the rubbish collectors. Get the idea?

When you go out and shoot you should not attempt to introduce all the shooting and directing skills in one big bang. You will come back with unbroadcastable rubbish and then you will get down-hearted. Gradually, you should turn the camera's automatic functions to manual. Bit by bit, you should introduce more esoteric shots to your repertoire. Don't rush it. Shoot lots of things over many days and get feedback from people who know how to shoot and direct every time you return. They will tell you how you can improve your framing and selection of shots.

The chances are the tape you use will last 30 minutes. If your camcorder can produce colour bars (like you see on a testcard) then flick the switch and record about 30 seconds at the very start of the tape. It's good practice to do this before you head off. Two-person crews using a portable audio mixer will also record a piercing tone at this point. Camcorders generally cannot do this on their own. The reason you record bars (and a tone) is that it gives the editor the chance to fiddle with the nobs in the edit suite to make sure the editing equipment is optimised for your tape.

When you are filming, as a rule you should shoot about twice as many CU shots as MS and WS put together. You will realise why when we come to editing. Also, make sure that you shoot each action or element three or four different ways. You have XWS, WS, MS, CU and XCU to choose from.

Remember to shoot as many different elements as possible. This piece will last a minute. Let's say you use three shots of each element (the standard WS, MS and CU) and each shot stays on screen for four seconds. This means you need 15 shots of five different elements to barely cover your piece. However, some of them may not work so allow yourself some flexibility and get more. Look for at least ten different elements for this piece. Ultimately you should be looking for at least 20 different elements shot three different ways. That will allow you to cover most stories.

And finally, for goodness sake, hold your shots for 20 seconds. It will spare you a lot of frustration.

The scurrying clouds

What is it?

A super XWS of stormy clouds tumbling and rolling across the screen in a scene to rival the most artistic of Australian outback movies. As we watch, the rays of sunshine split the sky like the hand of God itself igniting patches of the planet Earth in a warm glow.

How do I do it?

This one is really simple. All it takes is a bit of time, and you have to choose a day with clouds. The best kind of skies are about half clouds and half deep blue. Just like you get following a tremendous downpour in spring. Get your wide-angle adaptor and set your camera up as close to the ground as you can. Frame the entire frame with the panorama in front of you including a large patch of the sky and the ground. Then set your longest tape on record and go off and get yourself a nice hot cup of coffee. When you return, assuming nobody has stolen your camera, you will have a tape full of what appears to be nothing much. Until you speed it up in the edit suite. Not only will you capture the clouds rushing across the sky but you will also record their shadows on the landscape. If you're shooting in a park or in the country, the result will be devastating.

Now you are back

That was quick. You now have a tape packed full of WS, MS and CU shots that will all cut beautifully together to produce flowing sequences. Great!

But before you go leaping and jumping towards the chapter on editing, you have some housekeeping to do. First, you need to write some details on the tape box and the tape label. You cannot tell tapes apart by taste or smell and you need some way of identifying the one you want from the huge piles that litter your desk. You want to write the following basic information on both the tape and its box:

(1) Your name
(2) Shooting date
(3) Shooting location
(4) Story name
(5) Tape number in sequence

The tape number refers to the total number of tapes you used for a particular story. Let's say you used three tapes on a story. They are all marked up identically except for the number which goes '1 of 3', '2 of 3', and '3 of 3'. Meaning, for instance, this is the second tape from a set of three.

You may also wish to write some keywords on the box that refer to elements that may be needed at a later date in another package. Pictures of celebrities or specific locations spring to mind. Now you can sleep easy at night knowing you will be able to find your tape from amongst the heap.

The next thing you need to do is log your tape. It's no use knowing that you have shot an exquisite CU of a bee sucking nectar from the head of a chrysanthemum if you cannot find it when you are under pressure in an edit. It takes time to fast forward and rewind through tapes and, as anyone with a VHS knows, it takes a lot longer if you are picture-searching at the same time. So what you need to do is go through your tape and write down on paper every single usable shot you have taken. But first you will need to learn about timecodes.

Tracks and timecodes

Let's stop for a moment and look at what is stored on the videotape you have just brought back. First, pictures. This is obvious. When you play the tape you can see them. Second, sound. The microphone on the camera will have recorded whatever sound was going on while you were shooting – not very well but it will be there. In fact, it records two tracks of sound. If you have just one microphone built into your camera these tracks will both be the same. In effect, it is mono sound. If you have two microphones built-in then your camera's manual will claim it is recording stereo sound (though quite how two identical poor-quality microphones mounted right next to each other will produce the kind of sound people usually associate with stereo is beyond me).

In addition to the video and two sound tracks, your camera will have recorded some sort of timecode. This is much more than a mere tape counter. It uniquely identifies every single thing on your tape down to the frame.

You know how you can see the individual still pictures on cine film when you hold it up to the window. Video is no different. The moving picture you see is still made up of many frozen frames. They just happen to be recorded electronically on tape, that's all. PAL and SECAM television is made up of 25 frames per second. NTSC has 30 frames every second. Each one of those frames has its own timecode and it is written like this: 01.15.45.11 meaning the 11th frame of the 45th second of the 15th minute of the first hour.

Hopefully, your camcorder will let you set the timecode to some sort of accepted house style before you go out. House styles usually dictate what the hours field should read. As most people use 30-minute tapes, the hours field is not really used so most house styles invariably set it to 01 or 10. If you use more than one tape on the shoot then the second tape's timecode can be set to read 02 or 20 and so on.

It is also good practice to start the timecode at something like 00.59.30.00 or 09.59.30.00 so there will be 30 seconds of grace at the beginning of the tape before the first shot kicks off at 01.00.00.00 or 10.00.00.00. You will find out why you need that when you come to edit. You can fill that 30 seconds with the colour bars if you like.

Now you know all about timecodes, you can rewind your tape and fill in a logging sheet. This has three columns: timecode, type of shot, description. You do not need to log shots that did not work because nobody will be interested in them. And you do not need to mark them up to the frame. Seconds will be quite accurate enough. The example below shows how you should write this out.

Timecode	*Type of shot*	*Description*
01.00.00	WS	The main room
01.00.30	MS	A clock on the wall
01.00.50	CU	The pendulum swinging
01.01.10	CU	Cogs turning
01.01.30	CU	Second hand going from 45 to 10
01.01.55	MS	The players at the table

At this stage you will realise that not all MS shots are the same and you have to decide on some degree of relativity. So one sequence's MS may be another's WS.

Evaluating your work

When you are viewing your rushes, be really critical. This is where you learn what does and doesn't work. Ideally, there will be somebody who has had training who can talk this through with you each and every time. Ask yourself:

(1) Are you indeed holding your shots?
(2) Is the camera rock-steady?
(3) Are the pictures in focus and correctly exposed?
(4) What is the ratio of WS to MS to CU shots?
(5) Where are the transition points that will make easy cuts?
(6) What is the framing like on each shot?
(7) Do the pictures really look like the kind of television you see at home?
(8) If not, why not?

After you have completed your tape log, take your tape to the library and register it with all these details intact. The librarian will love you.

The main points

- Choose stories that immediately suggest appropriate pictures.
- Shoot at least twice as many CU shots as MS and WS shots put together.
- Shoot each element three or four different ways.
- Mark up the tape and the tape box with your name, the date, the location, the story name and the tape number.
- Log your tape as soon as you get back, listing each shot with a timecode, shot type and description.

6 Scriptwriting: the words

A station I once worked at hired what's called a TV Doctor. The job involves swanning in and constructively criticising all the station's output. Programmes, promos, trails and the like. Our doctor did his job with scant regard to the conditions under which the programmes were being produced. He was also paid a whole lot more than any of us who were actually doing the work.

He talked a lot about scriptwriting. And he had a set of rules that governed the sentence construction of television scripts. Most of them were too vague or ridiculous to go into here. But chief among them was threes. He liked threes.

Threes as in: over-paid, over-sexed and over-here. He wanted to see them in television scripts. You see, threes produce a nice rhythmic balance to the sentence when it is read out. The third point in a list of three always sounds like a final definitive statement. It makes you want to stand up and applaud. Politicians use them a lot at political conferences.

But rules like this are not appropriate for television scripts. They are far more suitable for some rabble-rousing speech delivered by a party leader to the faithful, where the chief aim is to show conviction rather than substance. In television, our chief aim is to communicate. And to communicate as simply and effectively as possible.

None of us would have minded so much if the TV doctor hadn't then shown us a VHS of a feature he had scripted about some mountains out west in the US. The pictures were superb. Great monoliths bursting through the desert. Deep red towers thrusting high into the sky. Majestic old timers of America's history. The trouble was the voice-over told us how fabulous they were. (It was a kind of James Earl Jones voice.) Everything we were looking at was described in the voice-over. Pardon me, but one reason you have pictures in television is so you don't have to use adjectives to describe what you are looking at. You can see it. It's there on screen.

What a script is for

Scripts in television are unlike stories in newspapers or on the radio. They do not have to go into every last detail. In television, the pictures usually tell the main story. A newspaper may mention someone's age, hair colour and the street they live in. On television, viewers can see roughly how old they are and can read the street sign. You do not have to say 'this blonde 14-year-old schoolgirl lives in Aardvark Street' if you have a shot of her in her uniform walking past the sign. The pictures speak for themselves. The script has other jobs to do:

- It can explain what we are looking at and why.
- It can impart background information that you cannot shoot.
- It can tell a parallel story that is connected to the pictures.

The script should never simply describe the scene being shown but it should always refer to it in some way. It's called matching the pictures to the words.

In the case of our training athlete, the pictures strictly show working out on the step machine. That topic, in itself, would not make a particularly interesting voice-over. But the shots can be used to cover a script that talks about the athlete's training regime, his health, diet, obsession with fitness – any number of related subjects. Conversely, the step machine pictures would not be suitable to cover a script that goes on about about the athlete's social life, his interest in the Internet or his family history. Unless, of course, the step machine impacted upon these things in some relevant way.

Writing scripts for television packages is a lot freer than writing news in print. Print has a set formula for the construction of a news story. The crux is told in one line right at the top. The rest can be cut from the bottom up without losing any sense. In television, you can make the assumption that the viewer will stick around for the duration of your package – if you have shot, directed and scripted it well enough.

What you are doing in your television package is telling a tale. This means that each line of the script does not have to be some hard fact. It also means your script should be structured with a beginning, a middle and an end. This gives you a broad licence to be, dare I use the word, creative when writing scripts for television. But it does not mean that you can make things up. It simply allows you to use your imagination to come up with novel ways of putting your story across.

Where the information comes from

Like any journalist working in any medium you have your information sources. These may be cuttings from newspapers, magazines or wire services that have already done a lot of the research for you. Or you may have a comprehensive library packed full of background information on every subject under the sun. But more likely, you will have to do your own research with the people you are shooting. And this means talking to them – extensively.

You can get a lot of the information you need for your script in advance over the phone. When you are on location, you will be busy shooting and directing so the more factual research you do before you head off the better. But scripts should also react to events that happen on the shoot. So don't be so rigid with the structure of your script that you cannot include good televisual events that spontaneously happen on location.

If you manage to capture your subject being thrown from their horse, for instance, make sure you include a line about everything not going according to plan. Indeed, if you are sent to make a package on a breaking news story you will be almost totally reliant on what you pick up from talking to the people on location. You simply need to make sure that when you return from the shoot you are confident you have enough information and ideas to write your script. Eventually, with practice, you will be structuring and writing the script in your head while you are shooting and directing and chatting to the subjects.

The Point Of View shot

What is it?

The killer in the movie Halloween is hiding behind the hedge watching purity itself in the form of Jamie Lee Curtis walking home from school. She looks up. We cut to a shaky shot of Jamie that ducks behind the hedge. Clearly, this is the murderer's Point Of View. Too elaborate for videojournalists? A colleague was doing a story about wild boar in one of the royal parks. We saw a MS of them snuffling through the undergrowth. We then cut to a POV shot that moved around in the same way as the pigs. It worked really well.

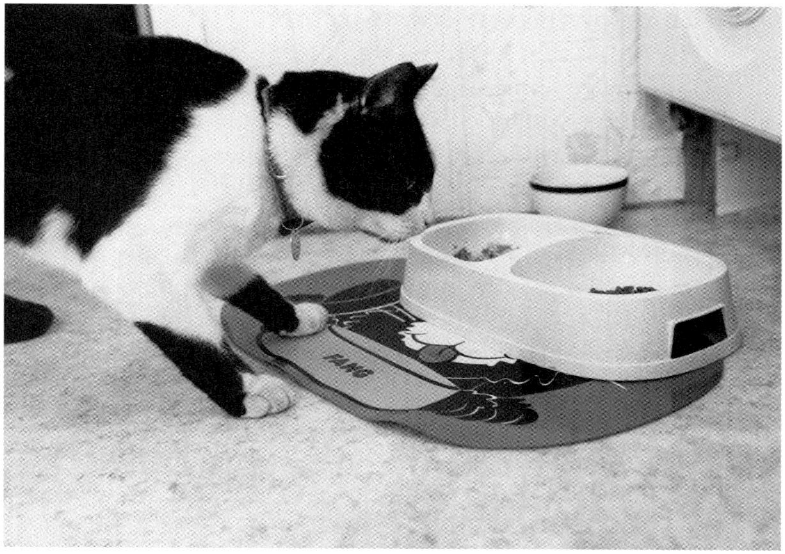

First, establish through who's eyes we are about to be looking

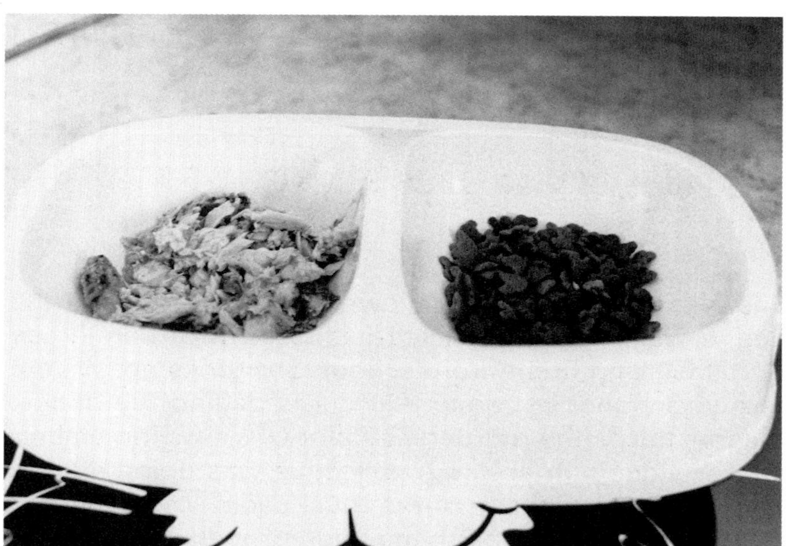

Then cut to your POV shot

How do I do it?

The fundamental important thing is to establish exactly who's point of view we are looking from. All you need is a MS of someone or something clearly showing them looking around or moving their head. The POV shot you follow it with emulates the movement we have just seen. Then it's obvious what's going on.

It's probably a bit contrived for news but a shot like this will certainly liven up your features and documentaries. Just remember to keep the movement of your camera as close as possible to natural eye movement. Resist the temptation to swoop around because nobody's eyes do that. Instead, hold the shot still on one thing and then move it quickly to another thing.

Introducing the story

We have already established that the first few lines of a printed news story pretty much tell you what the story is all about. The same is true in television news. It is called the intro.

This is what the news presenter reads off the autocue just before the package is shown. Often you can see the presenter's eyes moving and sometimes they read it like they have never seen the text before in their lives. That is probably because they haven't.

It always makes sense for you, the reporter, to write the intro. After all, you know what the story is about. You know what is in the script. It is usually only a few sentences long and ends with 'Anastasia Chappaquiddick reports', or whatever your name is. Alternatively, if you are used to writing printed news stories, the intro can be the first few lines of your story.

One hint about writing intros: if any of you, at any time, on any station, use the phrase 'with a difference' in an intro (as in 'Anastasia Chappaquiddick reports on the restaurant with a difference') then prepare for the hand of whichever God you believe in to descend and smite you dead. What, exactly, is this difference? Are they all nude? Is the menu exclusively in Polish? Let me tell you: 'with a difference' is meaningless. Everything is different in some way. My cat is different to the next black and white cat, not that the casual observer would notice it. If you can't think of any other way of inflaming the viewer's interest than claiming your subject is something 'with a difference' then you should seriously question why you are doing the story. Maybe it means you shouldn't be.

Laying out the script

Somehow you have to bring the pictures and the words together. Thankfully, there is an accepted format for writing television scripts. Most newsrooms have computers that are specifically designed for this purpose. You write your television script double-spaced in two columns with the words on the right and the pictures on the left, as in the example below.

	(INTRO)
	A player's been accused of
	cheating at the World Board
	Game Championships being
	held this week in Earl's Court.
	More than 300 players from 34
	countries are taking part . . .
	and some of them take it very
	seriously indeed. Anastasia
	Chappaquiddick reports.
	(BEGINS)
CU Coffee water dripping	
MS The coffee machine	This coffee pot's been filled
steaming	eight times in the last hour.

	It may only be a game.
WS Establishing the layout	But don't tell that to the players.
CU Sweat on player's forehead	Up for grabs is 5,000 pounds prize money and title Board Game Champion of the World.
MS Players round the table	It's usually an amicable affair. But this year, something's gone wrong.
CU Player reaching for the dice	The semi-finals are normally through by the second day.
CU Player cringing	This game dragged on for four.
CU A hand shaking the dice	That's when the organisers decided to investigate.
MS The clock on the wall	They now say they've found evidence
CU The pendulum swinging	that at least one of the players has been cheating.
CU Second hand reaching 12	
CU The dice crash to the table	
XCU The numbers on the dice	The claim has caused outrage in the world of board games.

CU Player scowling	And some people are calling for the executive committee to resign.

You will notice that the intro tells the story but doesn't, ahem, give the game away by revealing too many details. Similarly, the voice-over does not rattle on the whole time. The tension-building clock sequence can stand alone. More on working with pictures and words in the next chapter.

The guidelines

As a videojournalist, you will be recording the voice-over that accompanies your piece so the most important thing as far as you are concerned is that you can read the script without tripping over yourself or swallowing words. Rather than handing out complicated rules that govern the construction of sentences in television scripts, here are a few guidelines that will simply make life easier for you. Take a deep breath, there's a lot of them.

- *Keep sentences short and simple* – otherwise you will run out of breath before the end producing an unnatural break.
- *Avoid flowery language* – you are not writing an A Level English essay or entering a creative writing competition.
- *Cut out unnecessary words* – if you can use one word instead of five then do so because the sentence will be easier to understand.
- *Don't use subordinate clauses* – believe me, by the time you continue your original sentence the viewer will have forgotten what you are referring to.
- *Avoid starting consecutive sentences with the same word* – particularly watch out for the word 'the'. Rearrange sentences if words become bunched together.
- *Avoid using the same word twice in a sentence* – that sentence, for instance, would sound a lot worse if it read 'avoid using the same word twice in the same sentence'.
- *Avoid starting consecutive words with the same sound or letter* – try saying 'the theatre thought three thousand thespians . . .' when you are under pressure of a deadline.

- *Don't start a new sentence with the same word you finished the last one* – it sounds odd, that's all.
- *Don't be pompous* – most people do not say things like 'with which' or 'to whom' in their everyday speech so don't put them in your script. If you don't want to end a sentence with a preposition then rewrite it in a simpler way.
- *Don't be too technical* – simplify concepts as much as possible. Viewers will buy a magazine if they want really detailed specialist information.
- *Assume your audience knows nothing* – you know lots about the story. The viewer does not. You have to provide the context for your piece so they understand its implications.
- *Never use 'remains to be seen'* – as in 'it remains to be seen whether'. This is usually followed by some absurd pontificating on the part of the scriptwriter that is presumably intended to be poignant and thought-provoking but instead comes across as being crass and shallow because no conclusions are drawn. It's a horrible cliché.

Recording the voice-over

Before you record your script, read it through out loud. You will be surprised what you come across and you'll appreciate why I've given you all those handy hints. If you find something difficult to say when you are practising you can be sure you will completely foul it up when the tape is rolling.

Watch out particularly for odd names of people and places. Write them out phonetically and practice them on their own and within the sentence. You don't want to get tripped up by them while recording.

You will also find typing errors that have changed words. Maybe the spell-checker didn't pick it up but you can be sure the viewer will if you say 'ministers are on Brussels today' instead of 'ministers are in Brussels today'.

Make sure you vary the tone and pitch of your voice to make the commentary sound interesting. If you come across as bored then the viewer will assume the piece is indeed boring. Go through the script and underline important words that require particular emphasis. This will help you get the right flow for the voice-over. Put question marks and other annotations on the paper to indicate inflection if you want to. But take care you don't end up sounding like an over-zealous radio DJ.

Finally, remember there are full stops in the script and this is not some kind of race. You don't get a prize for finishing it first. Speak slowly, clearly and naturally.

That's really all there is to being a voice-over artist.

The main points

- The main aim of a script is to tell a story.
- Write an intro for the news presenter to read.
- Keep your sentences simple and clear.
- Let the pictures fill in the detail such as hair colour or street names.
- Make sure the script refers in some way to the pictures.
- Read your script out loud before you record the voice-over.

7 Scriptwriting: the pictures

You may feel that we have been spending far too much time concentrating on the visual aspects of television rather than the journalistic side of it. One chapter on scriptwriting compared to countless chapters on shooting and directing.

Well, here's the rub. Pictures are far more important than words for television. It is much easier to make up a convincing and plausible script to cover some fab pictures you have of a riot than to connive together a bandy collection of stills, graphics and aging stock footage to cover what should be a print or radio story.

The woeful lack of appropriate pictures every time a story on the economy is done on the television news merely serves to highlight the importance of putting the emphasis in news bulletins on angles that are inherently televisual.

For instance, instead of showing pictures of bank clerks under a voice-over talking about abstract concepts like rising interest rates, what about doing a story on someone being evicted from their home because they cannot afford their mortgage. It's a matter of coming up with the television angle.

When you come to write the script, you will immediately know whether you have been successful because the pictures and the words will flow together. You will not be left with large tranches of script lying bare without pictures because nothing suggested itself.

The chicken or the egg?

So do the pictures or the words come first? Well, it depends. The real skill in making cracking television is to choreograph the voice-over, the pictures and any important sound you may have captured on tape.

Let's say you are doing a feature on someone's search for a childhood sweetheart they haven't seen for 60 years. You have a poignant punchline coming up and you don't want to give it away beforehand in the voice-over. Read through the script below.

Pictures	Words
WS Arthur walking up garden path	The trail led to this house . . . in a quiet street . . . in a forgotten suburb.
CU Arthur knocking on door	
MS Door opening and Arthur talking	
	But Arthur's quest was not quite over.
XCU Occupant shaking head	
	Although Alice had lived here for 20 years, she was no longer there.
MS Arthur head-on looking up	
CU Cemetery sign	Arthur had found her . . . but it was three years too late.
MS Arthur approaching a grave	
CU Arthur crouching by grave	

You see how the pictures and the words interact to tell the whole story. There is a lot that does not have to be said and the piece is stronger for it. You are picking out details with the camera and alluding to them in the script without driving the points home with a sledgehammer.

If you are making breaking-news packages you will probably not have the time or the opportunity to craft your pieces with such skill. You will feel triumphant if you manage to introduce some hint of subtlety into your packages. A colleague tells proudly of a piece he did on a missing dog (there must have been more to the story than that to have justified covering it on a regional news programme but it escapes me for the moment). The piece ended on a lone shot of the lead hanging up on its own, abandoned and unused. It was a tear-filled moment. Apparently. He likes to tell that tale because it was a news story that became rather more than the usual catalogue of pictures, any pictures, slapped on top of a voice-over.

If you are making features or documentaries, then not only will you have the time to make more integrated television but it will be expected of you. As a videojournalist, you will have an immediate advantage when it comes to tying your pictures to the script because you shot them. So often in conventional productions the first question the producer asks when returning from a shoot is: 'What shots have we got then?' He or she then scurries off to view the tapes and start making decisions.

You, however, know exactly what shots you have got on the tape. In fact, the advantage goes further than this. Not only have you seen the exact shots you took while filming but they were exact shots based on informed decisions about how you were going to use them in the finished piece. In effect, every last frame of your shooting has been precisely tailored to the script you will write that will ultimately tie them together. Usually, only movie directors have that degree of precise control.

The perfect rack focus

What is it?

This is something your eye does all the time, and it is a camera technique that is often used in American drama. The picture has two elements of interest. Usually these are people. One is in the foreground and the other is in the background. The one who is talking is in focus while the other is blurred. As soon as the

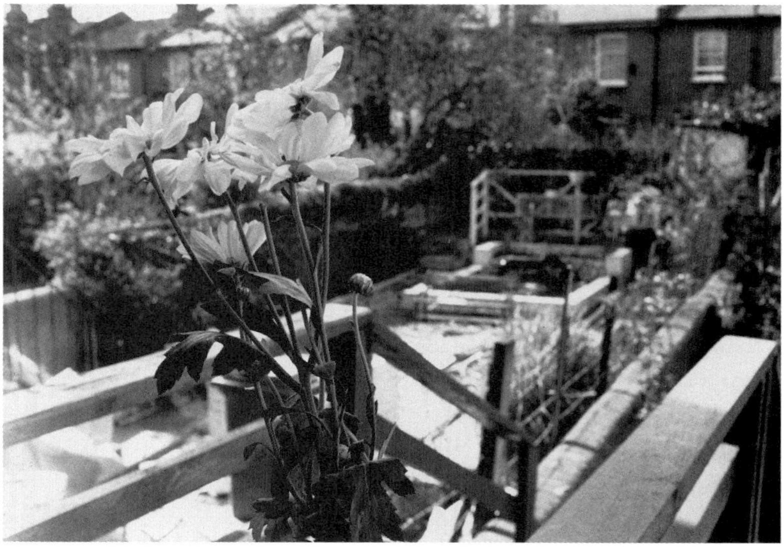

At first, the foreground interest is sharp but then you carefully turn the focusing ring . . .

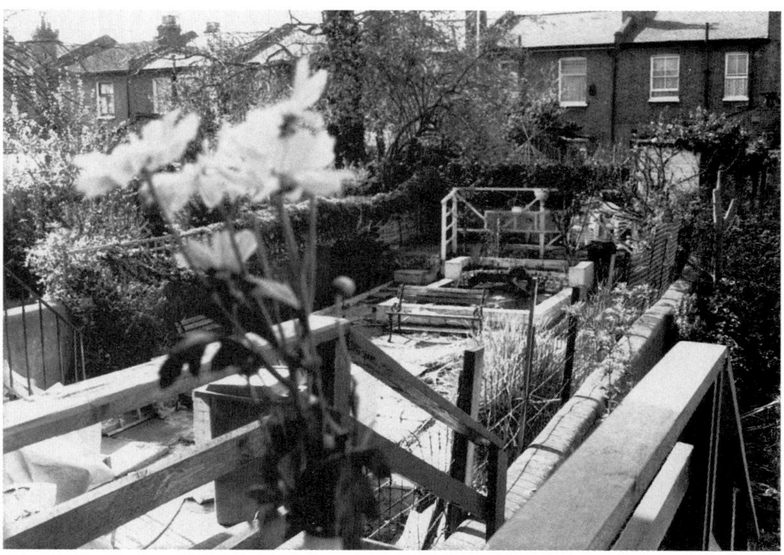

. . . and the background magically becomes clearer

line is finished, the other person starts to talk and magically they are now in focus and the original talker is blurred. You can do the same kind of thing in your stories with a flower that's in front of a view or a fence that's in front of a waste tip. It's a very slick way of focusing the viewer's attention on different things without cutting.

How do I do it?

Zoom right in. All the way. Then, move right back and put the camera on the tripod. Technically, what you have just done is reduce the depth of field for focusing to very little. With the camera rock-steady, frame the picture so that your foreground interest is on one side and the background interest is on the other. When you adjust the focus, you will see that either element can be sharp but never both at the same time. Start recording and hold the shot for the usual 20 seconds. Then, doing your level best to avoid shaking the camera, adjust the focus so the other element becomes sharp and hold the shot for another 20 seconds or so. You've got it.

Writing your script

What you now have to do is to write a complete marked-up script for the rushes you have already taken. Take a look at the tape log. You have already seen the pictures twice – once while you were shooting them and once when you logged them. Write your script in the two-column format introduced in the previous chapter, with the words on the right and the picture descriptions on the left. You should also write an intro at the top of the script on the right-hand side. Check all the handy hints listed in the previous chapter and keep your shot descriptions fairly short. The timecode will uniquely identify the shot on the tape. At the moment we are not worrying about sound. We simply want good writing and pictures that evidently tie in. The script below shows you what I mean.

	(INTRO)
	Scores of seagulls covered in oil have been found washed up on beaches along the south coast around Bognor Regis.
	Rescuers say it's the worst wildlife disaster in the area for nearly 30 years.
	Fabiana Francisco reports.
	(BEGIN)
10.04.25 CU Struggling bird	
	These birds are the lucky ones.
10.04.45 MS Birds on beach	The tide's carried them all the way to the shore.
10.03.55 WS Rescuers picking birds off beach	It's a simple matter for rescuers to pick them off the beach and take them away to be cleaned.
	But out there, are potentially
10.13.05 CU Wildlife entrance sign	hundreds more.

10.13.55 MS Person washing bird	This wildlife hospital usually deals with less than ten oiled birds every month. Over the last day, they've cleaned up 60.
10.13.30 CU Squeezing washing liquid 10.14.20 WS Washing bird	Rescuers are working overtime and volunteers have been drafted in to cope with the disaster.
10.14.55 MS Clean bird being put in pen	These birds will get released as soon as the water's clean again.
10.15.25 CU Clean bird	But the coastguard says it doesn't know who caused the slick.
10.02.20 WS Channel and ships	They're blaming one of the 300 ships that pass through the Channel every day.
10.03.20 MS Rescuers picking up birds	One of them must have illegally discharged the oil when it was foggy.

10.05.05 CU Birds being carried to van	The worst previous disaster was in 1967 when more than 100,000 tonnes of crude oil spilled from the *Torrey Canyon*.
10.05.45 CU Birds being put in van	All the rescuers can do now is hope the high winds and choppy sea will get rid of the oil overnight.
10.07.05 WS Van drives off, sun sets over sea	This is Fabiana Francisco for south-east news in West Sussex.

Here are a few things to note:

(1) The pictures are directly referred to in the first sentence of the voice-over.
(2) The change of location is also tagged in the script, but we don't need to say the name of the wildlife hospital because there's a shot of the entrance sign.
(3) The voice-over gives a few easy figures about the number of birds they usually treat to put the disaster in perspective.
(4) The CU of the washing-up bottle shows how the birds are cleaned so we don't need to spell it out in words.
(5) The shots of clean birds in a pen complete the whole bird-washing sequence so viewers know exactly what happens.
(6) For legal reasons, we only take a WS of ships in the Channel when talking about who may have done the evil deed because we don't want a libel suit falling on our desk.
(7) The closing shots show the work is still going on to imply the story is too.

If your script looks something like this one then you will be well on your way. At this stage it is important you get your script checked by the script editor or news editor or whoever it is you report to. If you report to nobody then get someone else to read it through because this is really the last point that problems can be ironed out before you commit yourself.

After getting the final okay, the next thing you need to do is to record your voice-over on tape or disk. Then, with your fully marked-up script in one hand and a tape of your voice-over in another, you're ready to edit.

The main points

- Good television is a marriage between the voice-over and the pictures.
- Let your pictures and words work together to tell the whole story.
- Think of ways the pictures can provide a punchline to your script.
- Get your final script approved before recording the voice-over.

8 Editing

When I started working in television I was told that the best way to learn how to make it was to edit somebody else's pictures. That way, you find out exactly what shots you need to tell a story without the terrible expense of actually going out on a shoot. You also learn to berate the person who was in charge of the shoot because of all the shots you desperately need that they missed. So you pledge to do better and think ahead when you are the top dog on location.

For now, however, you're not going to edit somebody else's rushes – you are going to edit your own. But for our purposes, it makes no difference. You see, we can go on about the importance of ordering sequences, holding shots for 20 seconds, and getting lots of CU shots until the tape itself degrades, but unless you have sat down in an edit suite and put the whole lot together you will never really understand the grammar of television.

At one station, I worked with a news cameraman who had never stepped foot in an edit suite. He had no idea how his shots were used. And, bizarrely, he took great exception to anyone else suggesting certain shots he should take while filming. On the few occasions I was lumbered with him, I found a lot of the shots were unusable. Not because they were poorly white-balanced or badly framed or shaking all over the place, but because they were not held for long enough or they did not cut. Even using conventional techniques, making television is not a series of single jobs. It is a collaborative process of interlocking skills that culminates in a single product. This is why so-called multi-skilled television production works so well. It is also why understanding what you do in the editing process is of singular importance to the production of high-quality television.

This is editing

Editing is not just about stripping out the waste from your rushes, or reducing the running time, or simply making what you shot look better. Editing is how the television programme comes alive and takes on its own personality. It is the stage at which things succeed gloriously or fail terribly. An elaborate sequence may seem to run well on the paper script but it can easily flop when put together. Conversely, a shot you had rejected at the paper stage may turn out to be the centrepiece of your package. When you edit, you discover why you cannot cut from one MS to another in the same scene. And why crossing the line looks so bad.

The editing process is where all the shots you took at different times on different days are brought together to make a single flowing piece of television. So you will need to get your mind around it.

As well as simply putting together your sequences, you can use editing to add life and vitality to your package. For instance, you are filming some sort of gathering where the speakers are droning on. You are not going to need everything the afterdinner speaker says, so in the downtime get shots of people sleeping in the audience. You can then be naughty and paste the shots of the sleeping audience over quotes from the speaker to make them appear a tad dull. Great fun.

Before we look at how to edit, let's remind ourselves of the elements we have got so far to make our package. First there are the pictures. Then there is the background sound that accompanies the pictures. Finally there is the voice-over. Thankfully, when editing, we have three different tracks to play with. One video and two audio – the three tracks we have previously identified on the video tape.

Now I know that most editing systems will enable you to play around with more than these three elements. But it is like driving. If you learn to drive using a manual car, driving an automatic will be a breeze. So, learn how to edit using these three tracks and you will be able to cope with anything.

What you are going to do is to record the pictures on the video track, the sound that accompanies the video on the first audio track, and your voice-over on the second audio track. This is the easiest set-up to learn and all that you need for now, since you are putting together a simple package here and not making a movie. Strictly speaking this does not necessarily produce a tape that can be broadcast. The two sound channels, which correspond to the left and right speakers on a stereo television, are not mixed. It would result in different sounds coming out of each speaker. But we are focusing on the concept more than the end product at this stage.

This book is not about how to become an editor. For the first time we are introducing another major player into our multi-skilled production process. As they say, there is only so much a person can do. And the independent eye of a good editor is very important.

The edit suite

Edit suites come in all shapes and sizes. The simplest is a two-machine edit suite. This has two video recorders. One to play on and one to record on. Between them is an edit controller. This device duplicates the control buttons like play and pause on the front panel of the video recorders making editing easier. Or an edit suite may simply look like a computer with a huge screen. And that's exactly what it is. Or it may look like mission control at NASA and have lots of computer screens, keyboards, video recorders and monitors.

However the edit suite looks, in front of it all is the editor who does all the complicated stuff like making sure your voice-over is not obscured by the background sound. Your job is to drive the edit and say what you want to go where. But that doesn't mean the editor is there as a technician. In the same way that a script editor finds holes in your writing, the video editor finds holes in your directing and can vastly improve on whatever vision you had in the first place. For simple pieces, you may be able to get by doing it yourself. But for anything that requires even so much as a hint of quality, get a good editor in. They are invaluable.

The editing process, then, becomes your first real collaboration. At this stage, do not become precious about your work. A good editor has seen many different packages made by many different people. You should be able to rely on their experience to make your package better.

To begin with they need precise details about your package. A spare printed script is a good start. The reason you mark the timecodes of all the shots on the script is so that the editor can find them quickly without your having to guide them through every frame of the rushes. Shots that you have put together into a sequence are invariably found around the same section of tape. A good editor will see these as they are spinning to the first picture you want and may make suggestions as to how the shots can be reordered so that a more flowing sequence is achieved. Listen and be flexible. At the end of the day you have the last word, but putting pictures together is an editor's job so they are bound to have some valid ideas. Consider them.

Very simply, the editor will probably start by putting your voice-over on audio track two. They will then rewind to the beginning and start laying the pictures on the video track and the background sound on audio track one, without disturbing the voice-over.

To lay down a shot, your editor enters three of the following timecodes on the edit controller:

- IN point on the play tape
- OUT point on the play tape
- IN point on the record tape
- OUT point on the record tape

The editor needs to know these points to the individual frame. The IN point on the play tape is the same timecode you put next to your shots on the left-hand side of the script (though it is not necessary to write them down to the frame). The IN point on the record tape is where you tell the editor you want that shot to go (usually 'next'). The OUT point on the play tape defines the end of the shot you want to use and the OUT point on the record tape says where it must finish. Anyone with a grasp of basic mathematics will realise that you do not need all four pieces of information to make the edit. The IN and OUT timecodes together on either the play or record tape determine the duration of the shot you are laying down. So if you define an IN and OUT point on the play tape you only need to specify the IN and OUT point on the record tape. And if you define the IN and OUT points on the record tape you only need to specify the IN or OUT point on the play tape. The edit controller is then clever enough to work out what is going on when the editor lays your shot on top of the voice-over ensuring it cuts with the previous picture.

This is a very simplistic way of looking at editing. The best way to really learn what goes on is to sit down and do it.

What to do

You are learning the basics of quality television production so you should not be overly ambitious with your pictures in the edit suite. If the transitions between shots are too obvious then they will distract from the story. There are only two basic functions you will need to master: cutting and mixing.

Cutting

This is simply chopping from one picture to the next. The transition will be seamless if your piece is well directed. Viewers will not even register that the shot has changed.

Mixing

This is when one picture dissolves off the screen as another appears. Mixes are often used in a horrible way in an attempt to give artistic direction to a dull string of pictures – in films of horticultural exhibitions and the like. The trick is to use the effect for a reason rather than as a means of getting yourself out of a directing hole. Mixes work best when they indicate the passage of time. For instance, you have a shot of two people walking down a long coastal path with bags on their backs. It mixes to an exterior shot of the Youth Hostel. Just as the path finally disappears, our heroes traipse wearily into view. That would work well.

The invisible edit

What is it?

The term refers to a cut that the viewer is largely unaware of. There are two basic types of invisible edit. The first allows you to salvage something cool out of what you thought was a ruined shot when some idiot walked in front of the camera. The second is rather more specialised and only occurs when there are stills photographers around.

How do I do it?

If someone walks in front of the camera the likelihood is that there will be one or two frames that are completely dark as they move from one side to the other. Cut on one of these frames in your edit and the viewer will barely register the change of scene. When stills photographers are flashing about all over the place there are likely to be similar frames of burnt out white. Again, you can cut on these. For extra value from these situations, cut from one invisible edit to another. The result looks very slick.

What to avoid

Just like the camera, edit suites offer inexperienced videojournalists a vast array of effects to tart up their packages. Resist them. Indiscriminately wiping between shots or breaking the picture into little boxes that spin off the corner of the screen was cool on *Top of the Pops* in the 1970s but we've seen it already. As ever, the rule is whatever you do on the screen has to be for a reason rather than for effect.

Now you can make television

It has taken half the book to get to the stage that you can make a simple television package. The point about the way we are making television is that all the stages and skills that have been covered so far are treated as one single process. That's multi-skilled television production at its most fundamental level.

Before you move on, you should make sure that you are completely up to speed with everything we have covered so far – that's everything from basic camera operation to framing to directing to scriptwriting. You only get good at making television by doing it and listening to the, hopefully constructive, criticism of people who can already do what you are trying to do. So shoot lots and package lots. Because if you haven't followed the story so far, you will go into complete information overload in the next few chapters.

The main points

- You will learn more about directing by editing than by any other way.
- Listen to suggestions from your editor because their whole job is pictures.
- Make sure each shot is fully-marked up on the script to avoid wasting time.
- Only use cuts and mixes for the transitions between shots.

9 Sound

Let me tell you about this . . . it was hilarious. I went to a huge trade exhibition at Earl's Court. The organisers had decided to pump money into operating their own television channel for the duration of the exhibition. Televisions were scattered throughout the place, with the idea that participants could watch the drivel while they sat down for a coffee. Or something. There was a nice big outside broadcast gallery packed with what looked like the whole entourage of producers, researchers, directors, vision mixers, camera and sound operators, runners and presenters. And lots of people flapped around appearing hot and bothered. That's how television is often made.

The station's output consisted of interviews conducted by a presenter on a set in the middle of the exhibition. The presenter and the interviewee sat on chairs and talked. This was the good bit. They both held their own hand microphones. And because of the degree of background noise, they had to hold them close to their faces. Not only were their mouths obscured but the whole thing looked exactly like they were singing a karaoke duet. I kept expecting to hear a rendition of 'I Just Called To Say I Loved You' come bursting through the speakers. You see, they were using the wrong microphones for the purpose. It was seriously odd. You do not want to get caught out like that.

Different kinds of microphones

In the greater scheme of things, recording broadcast-quality sound for packages is not quite as important as recording broadcast-quality pictures. You are scuppered if your pictures are bad. But you can always turn the background sound down so it cannot be heard when you come to edit your piece.

However, there are times when the sound is really important. Like when you are interviewing somebody. Or when you are recording a piece to camera. Or when you are doing a story on aircraft noise. Or if you are in a riot and there are people shouting and screaming around you.

And even if the sound is not of noticeable importance, remember what we said earlier about good television being the skilful choreography of many different elements. Cars driving past look odd if you cannot hear at least a quiet roar of the engine. Swans opening their beaks and shaking their necks appear mute if you cannot hear the bark. You want to make sure the sound you record can be used prominently in your packages – even if you cannot anticipate which sounds will be important at the outset.

The key to quality sound recording is choosing the right microphone.

Built-in microphones

Camcorders always have a built-in microphone. On professional cameras, it is usually a mono microphone that records exactly the same sound on both audio tracks on the tape. Other camcorders tend to have two microphones tightly packed together that claim to record stereo sound on the two audio tracks. Either way, the microphones on the cameras are rubbish.

Manufacturers may protest in disagreement but if the microphones on camcorders were good for anything other than recording hissy background noise why would there be such an array of alternatives to choose from? So use the built-in microphone for background noise. Do not rely on it for anything else.

Hand microphones

These are the ones that caused so much mirth at the Earl's Court exhibition. They are the sort used by reporters when they record pieces to camera because they give them something to do with their hands. They are also used by presenters interviewing people on the hoof and, of course, by karaoke singers.

The whole point about interviewing people with hand microphones is that you push it in their direction when you want them to speak. You are controlling the proceedings so you have the microphone. You do not have one each. As a videojournalist, you will mainly use the hand microphone to record pieces to camera.

Hand microphones are usually omnidirectional, which means they pick up sound from all around them. So as well as your voice they can pick up the cars driving past and the aircraft flying overhead. Watch out for this.

Boom microphones

These are what we see two-person crews using. The microphone is encased in a big fluffy thing that the sound operator carries on a pole and calls Dougal. Now, as a videojournalist, I defy you to use a boom microphone whilst using a camera. If you can contort yourself sufficiently to do it then you belong in a circus. However, the microphone inside the fluff is of vital use to the multi-skilled professional for the basic reason that it is unidirectional. This means it only records the sound that is directly in front of it. All the incidental whooping, shouting and roaring gets cut out. If you need to interview somebody on the fly then a unidirectional microphone is essential. You hoist the camcorder to your shoulder and use your right hand to stop and start the tape as normal. The unidirectional microphone goes in your left hand and, hey presto, you can wield it in any direction you like to capture the sound that you need. This works.

Unidirectional microphones are also useful if you are recording important voices near to some major source of noise like a motorway or an enraged crowd. You simply position yourself between the subject and the noise and make sure the microphone remains pointed away from the racket and towards your interviewee.

Clip microphones

Clip (or tie) microphones are small microphones that are attached by a clip to clothes somewhere under the chin. Usually, the wire is fed from the back to the front of the shirt or blouse. The microphone is then clipped to a loose flap of clothing, like a tie or lapel. Clip microphones are usually used in studios, where they are attached to a small transmitter in the wearer's pocket. This avoids any accidents or unscheduled humour caused by people tripping over wires. You will use them when you want to do formal interviews with stationary interviewees.

Sunrises and sunsets

What is it?

Pretty self-explanatory. From an XWS as it rises over the skyline of a city to a CU as it descends slowly into the ocean, the sun provides you with a superb opportunity for capturing a beautiful shot that can be used anywhere in your piece. Shots like these can make the most decrepit of scrapyards appear like the Hanging Gardens of Babylon.

How do I do it?

The natural light at sunrise and sunset is different to the rest of the day so select the filter on your camera that you normally use for artificial light indoors. Open the iris up as wide as it will go and white balance if there is enough light to do so. If there isn't, use an indoors setting or put it on automatic. Then frame your shot and think of the awards you will win.

Recording sound

The most important hot tip for recording broadcast-quality sound is: make sure you are actually recording it. Hopefully, you are using a camcorder that has some sort of sound meter. This will indicate the level of sound that is being recorded on each track. If it does not respond every time you tap the microphone then something is up.

The most likely cause of an unresponsive microphone is that the battery is flat. A lot of microphones have their own batteries and it is important to make sure they are have a strong charge. Remember to turn the microphone off when you pack it away at night otherwise you will spend a small fortune replacing the batteries. And in any case, always have a spare microphone battery in your kit ready and waiting for the moment it gives out just as you are about to interview the prime minister.

Even if the meters show that sound is coming into the camera, you should always check exactly what is being recorded by listening in using good-quality headphones. This will be your only option on most cheaper camcorders.

Setting the recording level

Hopefully, your camera gives you the option to set the recording level manually. This will allow you to boost the volume of quiet sounds and suppress the volume of loud noises so all your sound is recorded at roughly the same level. Just like the iris and the white balance, cameras are able to set the sound level automatically. If you are just recording background noise, you can safely leave the levels on automatic. It's when you start to record something important like a speech or birdsong that you need to take control.

Cutting out background noise

It is easy to forget about the background noise when you are interviewing somebody or recording a piece to camera. You may not have registered that a plane is flying overhead but you can be sure the microphone has. On automatic, the sound level will have dipped to take account of the plane and your voice will have disappeared with it. Controlling the sound level manually is one way to ensure you only record what you want.

If there is a lot of background noise, the first thing you should do is to try to get rid of it by closing the window or telling people to shut up or asking them to stop mowing the lawn for a moment. However, a lot of the time these circumstances are beyond your direct control. In which case you need to:

- Move the microphone as near as possible to the speaker's mouth.
- Ask the speaker to talk more loudly than normal.
- Switch the sound recording level to manual.
- Reduce the sensitivity until the background noise is acceptable.
- Check on the meter and the headphones that you can still hear the speaker talking.

Splitting the tracks

Another way to ensure you get broadcast-quality sound in difficult conditions is to split the audio tracks. We already know that camcorders record two high-quality audio tracks. Depending on your camera, you can split these so that they record different things.

For instance, you can have one audio track on automatic and one on manual but both recording from the same microphone. One of the tracks is bound to have got a clean version of what the interviewee was saying.

Or you can have one of the tracks recording sound from the built-in microphone and the other recording sound from the clip, hand or unidirectional microphone. This way, you will always have a backup if anything goes wrong with your primary sound source.

Or you can plug two clip microphones into the camcorder and record your questions on one track and the interviewee's answers on the other track. Both on automatic, both on manual. Or some sort of mix. The system is very flexible.

Cheaper camcorders are not usually able to record two different tracks on their own. All you have to do is to get an audio store to knock you up a lead that converts two mono microphone inputs into one stereo output. This plugs directly into the camcorder and you can do what you like with the tracks then.

Dealing with music

Music presents its own difficulties in the production of packages. It will be noticed even if it is just in the background. As a result you have to take great care when editing shots where music is involved.

Background music

First, see if the shop (or whatever it is) will turn the music off for the duration of your shoot. If not, then you have to record what is called a wild track that will replace all the background sound after you have edited your sequence together. The alternative is that the background sound will jump horribly like a scratched record every time you cut. The pictures for the wild track aren't important, so record at least a minute of continuous sound at the location and remember not to chat to anyone while you are doing this.

Music as a feature

Let's say you are doing a story about a string quartet. You need a good quality recording of a lengthy section of playing that can be used for a few seconds at a loud volume, when it will be focused on, and slightly longer as background music. You should keep the music to a minimum or you will have to start paying musician's organisations lots of money for the rights to broadcast their members' work. Position yourself back a bit where you can get a well-framed WS and where the music is balanced. Point the unidirectional microphone at the players and record about one minute's worth. Next, reframe and get lots of tight CU shots of fingers pressing the strings, bows scraping, heads nodding musically, feet tapping, that sort of thing. A skilful editor will be able to bung these shots convincingly on top of the WS. The music will remain continuous. And everyone will wonder how you managed a multi-camera shoot.

Adding sound

The most obvious extra sound you will include in one of your packages will be a quote from an interviewee. But that's not all. Any interesting sound that you captured on tape can be incorporated in some way. Birds singing, dogs barking or children playing can all be given prominence on the sound track and will add depth to your piece.

Just like the pictures and the voice-over, important sound has to be marked up and flagged in the script. It goes in the left-hand column with all the picture instructions.

Quotes

Important sections of speech, such as quotes or pieces to camera, are marked using the standard acronym SOT, meaning Sound On Tape. The example below shows you how it should look.

10.13.55 MS Person washing bird	This wildlife hospital usually deals with less than ten oiled birds every month.
10.13.30 CU Squeezing washing liquid	Over the last day, they've cleaned up 60.

10.14.20 WS Washing bird	Rescuers are working overtime and volunteers have been drafted in to cope with the disaster.
SOT	(SOT)
IN: 10.11.23 'We've never seen anything like it . . .'	
OUT: 10.11.43 '. . . save more lives.'	
10.14.55 MS Clean bird being put in pen	These birds will get released as soon as the water's clean again.

One thing that works really well is letting the previous picture run briefly over the start of the SOT before the speaker comes into vision. It works for two reasons. First because it is something that happens in real life. You hear somebody talking and then see them approaching. You hear a car's brakes squealing and then see it crash. So it's good to hear the SOT for a couple of seconds before cutting to the talking head. The other reason it works is that it smooths the transition between the different elements of your package and helps create that illusion of flowing continuity you are trying so hard to achieve. You can mark the point you wish the speaker to appear with IN VISION on the left-hand side of your script.

Important primary sound

If there is something distinct on the tape that you want to pick out, such as a bird singing or a car revving, then you use the acronym UPSOT (turn UP the Sound On Tape). This works best when we are already looking at the shot containing the UPSOT so, again, its inclusion appears seamless – as in the following example.

10.14.20 WS Washing bird	
10.14.55 MS Clean bird being put in pen	Rescuers are working overtime and volunteers have been drafted in to cope with the disaster.
10.15.25 CU Clean bird	These birds will get released as soon as the water's clean again.
10.15.29 UPSOT Bird crying	(UPSOT)
10.02.20 WS Channel and ships	But the coastguard says it doesn't know who caused the slick.
10.03.20 MS Rescuers picking up birds	They're blaming one of the 300 ships that pass through the Channel every day.

Important secondary sound

If there is something on the background sound that you want to be heard underneath the voice-over then use the acronym NATSOT (NATural Sound On Tape). This will make sure it has more prominence than the rest of the background sound in the package.

More than likely, you will want to turn up the NATSOT at some point during the shot if it is that important – as shown below.

10.14.20 WS Washing bird	
10.14.55 MS Clean bird being put in pen	Rescuers are working overtime and volunteers have been drafted in to cope with the disaster.
NATSOT Birds crying	
10.15.25 CU Clean bird	These birds will get released as soon as the water's clean again.
10.15.29 UPSOT Bird crying	(UPSOT)
10.02.20 WS Channel and ships	But the coastguard says it doesn't know who caused the slick.
10.03.20 MS Rescuers picking up birds	They're blaming one of the 300 ships that pass through the Channel every day.

The main points

- Identify the most suitable microphone for the situation.
- Built-in microphones should only be used for background noise.
- Hand microphones should be used for pieces to camera.
- Clip microphones should be used in controlled interview situations.
- Unidirectional microphones should be used in difficult sound conditions.
- Check the sound you are recording with headphones.
- Always set the recording level manually for important sound.
- Split the two sound tracks to ensure you get one good one.
- Record at least one minute of 'wild track' sound if there is music playing.

10 Interviewing

'Just you?' It's a question thrown at every videojournalist at some point in their career and it's usually asked by people who have had experience of crews. How you handle it can pretty much determine the outcome of the piece you are making. A weak, thin smile followed by a tendency to drop things will result in a complete lack of confidence on the part of the subject and co-operation will be withdrawn. The alternative is that you brush the question aside and get on with the job of wiring them up and setting up the camera while chatting about last night's telly. You see, being a videojournalist is more than just getting on with the job at hand. You have another role to play.

Crews are impressive. They have huge cameras and large fluffy microphones. They carry lights and cables and ask where the power sockets are. They say things like 'turning over' and 'at speed'. They have an assistant who rushes out to the van to get that vital filter and then stands just off camera holding an enormous circle of white that reflects natural light on to the subject's face. They have a director who stands there pondering and asking them to reframe about a millimetre to the left. And they have a producer who set the whole thing up and wants to make sure everyone is happy because it's their neck that's on the line.

You have, well, 'just you'. And somehow you have to be just as impressive. In fact, you have to make sure that your subjects are not even given the opportunity to think that there may be anything odd about the fact you are there on your own. Your finished piece will speak for itself of course. But they can't be expected to know that.

Your job is to inspire confidence from the moment you meet your interviewee so you get all the co-operation and help and quotes that you need to make your package. This is achieved very simply. Know your equipment inside out. If they think you know what you are doing when you are framing, setting the sound levels and putting the microphone on them then they will think you know what you are doing with the story and their quotes. So they will help you.

Setting up

We are going to assume that you have a certain degree of control over the circumstances surrounding your interview. Clearly, if you are outside a court waiting for a defendant to emerge or camped out on some celebrity's doorstep because they may or may not be getting married, you will have to take your chances with the best of them. Most of the time, though, you will have some decisions to make.

Choosing your location

The first thing you need to do is select your location. You want somewhere that is both appropriate to the story and interesting to look at. Let's say you are interviewing a managing director in her office. You enter the office and she is sitting behind a desk. That's it. The walls are corporate grey and they are lined with books. Not very interesting. However, outside is the real office where the real work is done. It is much bigger and there are lots of busy people doing office-type things. You will get a better shot with this activity as a backdrop than the pokey separate room. Or imagine you are interviewing a stablehand about his work. Much better to do the interview with the length of the stable in the background showing horses snorting and munching hay than to do it in the room he drinks his tea.

The point is that, as with every other shot you get, you should try to make the interview look interesting. This means giving the picture as much depth and life as possible. The only thing you have to look out for is that the activity in the background is not distracting and does not upstage the interviewee. A busy office is fine as long as two people do not start having a fight half-way through. That would be far more interesting than the interviewee no matter what is being said.

Companies often like interviews to be conducted in the reception area, in front of their massive corporate logo that is surrounded by an elaborate fountain. If it is indeed impressive then it can be justified. If it isn't, you should find somewhere more interesting.

Framing your interviewee

Having chosen your location, you need to think about framing. The interviewee is going to be talking to you. They are not going to be looking into the camera. You need to frame them so that when they are looking at you, their line of sight cuts right across the middle of the picture. Position yourself so that if they are on the left of the frame, they are looking to the right of the camera and if they are on the right of the frame, they are looking to the left, as Figures 10.1 and 10.2 show. If you do not do this, they will look like they are about to fall off the side of the frame. As in Figure 10.3.

Set up the camera so that it is at roughly the same level as their head and sit yourself down so that they are not looking above or below the lens. If you get a shot that takes in everything from the top of the head to the bottom of the chest, you will have secured the classic interviewee pose.

Of course, you can do something far more avant garde than this and frame them so you are looking right up their nose or in their left ear and this should be encouraged if you are doing stories on struggling artists and the whole point is that it is a bit weird. But in most cases you do not want the viewer to be distracted by your bizarre shooting and so you will want to frame the subject conventionally.

If you are asking your interviewee a number of questions then you should slightly reframe the shot for each answer. This does not mean you change everything about the composition of your shot. You simply have to adjust the zoom to go from a WS through to a CU so that each shot is slightly different.

Figure 10.1
A classic pose with the interviewee on the left looking right

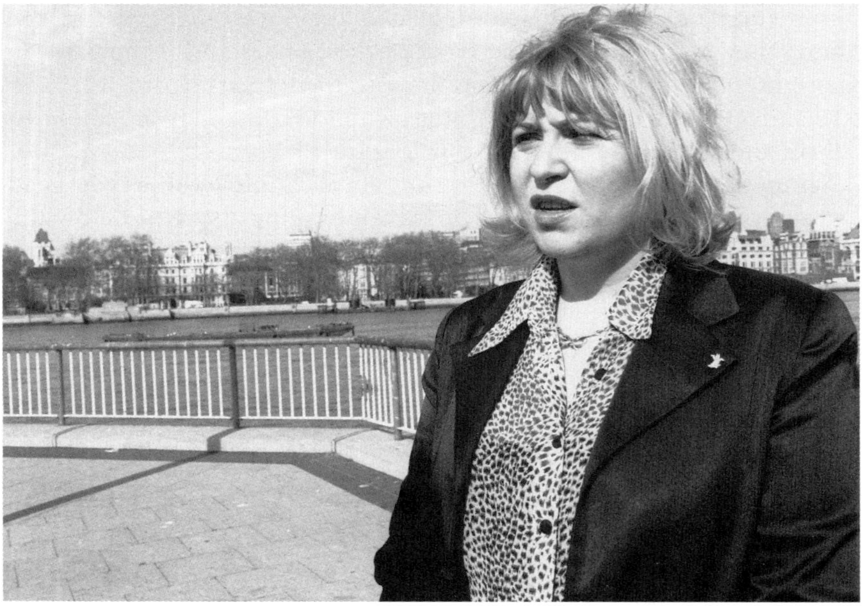

Figure 10.2
Or you can frame them on the right looking left

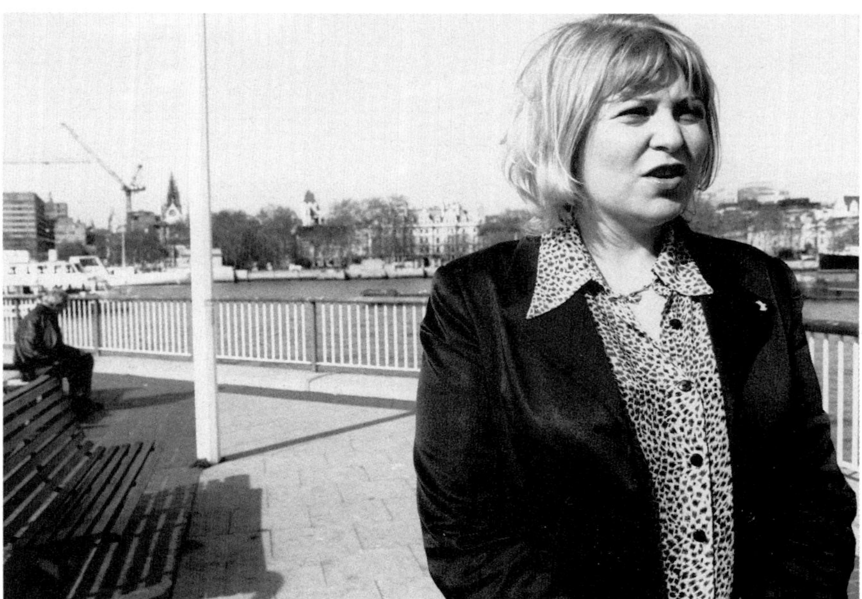

Figure 10.3
But if you frame them on the right looking right then it will look like they are falling off the side of the picture

You get two benefits from doing this. You should never use the same shot twice in a package. So if you want to use two quotes from a single person then it looks much better if one of those shots is a MS and the other is a CU. Second, by reframing each shot, shown in Figures 10.4, 10.5 and 10.6, you can cut between them very easily. This means you can use one sentence from one answer, say a MS, and then cut immediately to another sentence from another answer, say a CU. The result is a single answer that appears continuous.

If you have two different interviewees in your package then you should frame them so one looks to the left of the camera and the other looks to the right. There is no other reason for this than choreographing a nice balance of pictures in the story.

The only other point about framing is that CU shots are fine but XCU shots of faces tend to indicate that the subject is a shady character who needs to be scrutinised for signs of deception. Such shots, like Figure 10.7, are more at home in a Lucio Fulci horror film. So be careful about using them.

Checking the sound

You will probably be using a clip microphone when interviewing people. These are nice and discrete and are not distracting unless you

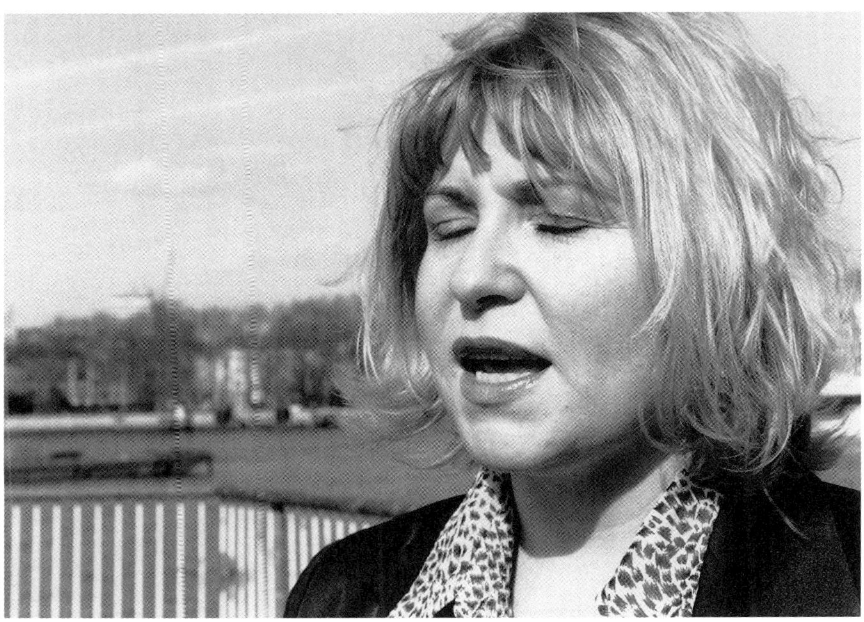

Figure 10.4
If you take a CU for the first question . . .

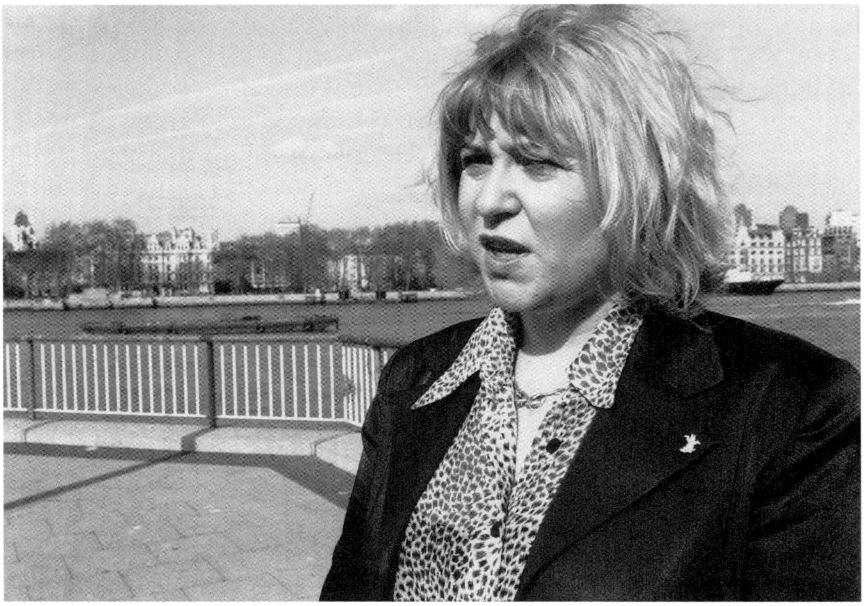

Figure 10.5
. . . then a MS for the second question . . .

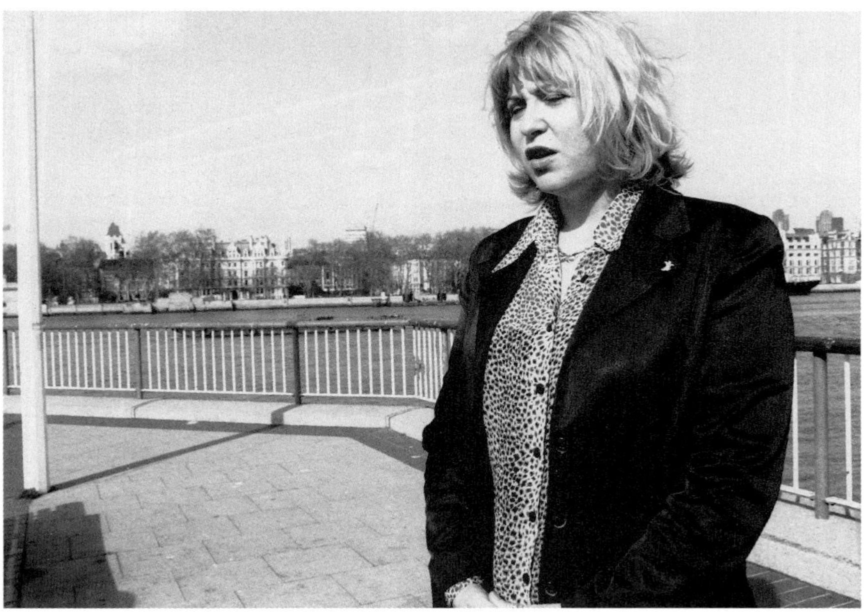

Figure 10.6
. . . and a WS for the third question then you can easily cut between them

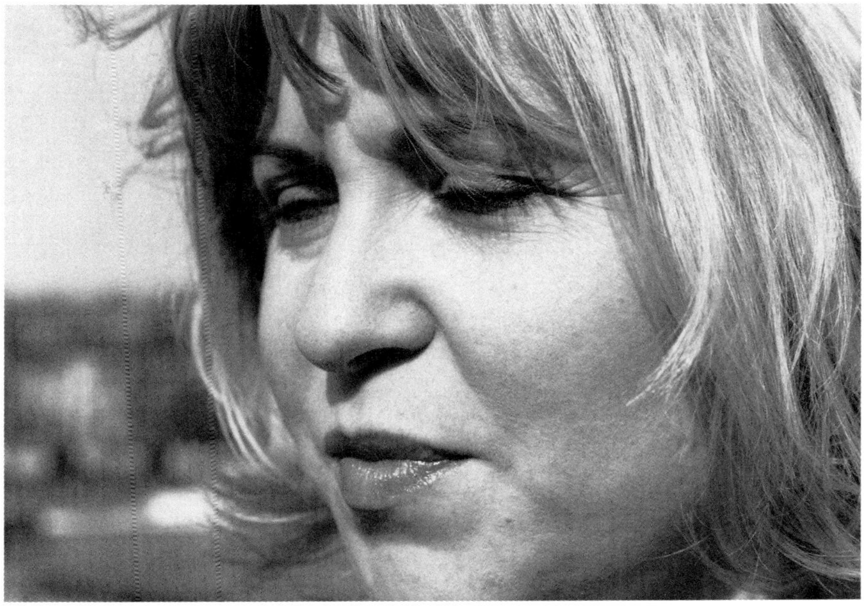

Figure 10.7 If your intention is to make somebody appear shifty and untrustworthy then by all means take an XCU

can see the wire snaking out of a hole further down the person's blouse or shirt.

If you can set the sound level manually then do so. Obviously the person you are interviewing needs to talk about something for you to do this so ask them questions about their role in the company or what they had for dinner last night. Once you have set the sound level, start the tape rolling and ask them to say and spell their full name and job title. This will help you in the edit when you come to put a title on the screen that says who they are. The only exception to this rule is extremely famous people, such as Joan Collins, who will think it odd that you do not know how to spell their name. Having done that, you are ready to roll.

Securing the quote

Most of the time you will simply be wanting a single quote lasting approximately 20 seconds from an interviewee. Much less and it will come across as too short. Much more and it will begin to look boring. If they are really that riveting then they should be interviewed in the studio rather than for a package.

Both you and the interviewee have the same aim. You want them to say something good that you can use in your package. They want to say something good that puts them in the best possible light. This is the sole purpose of the quote. The interview you do to secure that quote is not the interview you do to get information about the story. If this means sitting down for half an hour to discuss the issue together before turning the camera on then do it. There is no sense wasting tape even though it is cheap because you will only have to plough through it all again when you come to view your rushes.

Most people have seen television. They know that they will only have about 20 seconds to say their piece. As the scriptwriter, you know how you are going to introduce that quote into the package. So save everybody a lot of time and effort by agreeing the details with the interviewee beforehand and keeping the actual on-screen interview to two or three questions. You will know which answer you will use as soon as you hear it.

The other advantage of talking the process through with an interviewee is that you can make sure you get usable answers. If they have never done television before then they will not know that you want a clean burst of speech that can stand alone. Just like the vox pops, we do not want to hear your voice asking the question. And we do not want the quote to begin with 'because'. Media-savvy interviewees will know that they should repeat the highlights of the question in the answer. You need to explain this to everyone else. For example, you can't use the following answer as a quote:

You: 'Why didn't you do anything about the leaks?'
Answer: 'Because we don't believe in witch-hunts . . .'

But you can use this:

You: 'Why didn't you do anything about leaks?'
Answer: 'We didn't do anything about the leaks because we don't believe in witch-hunts . . .'

Difficult interviewees

At some point you will need to get a quote from somebody who clearly does not want to be there. They may give you all the time you want or they may only give you a few minutes. They may, even, remove the microphone and walk out as soon as they have said their piece.

The prisoner interviewee

What is it?

A lone soul sits bathed in light while all around is darkness. It looks like they are under interrogation. They blink slightly and speak. They're an interviewee giving you a quote and they look much more interesting and intriguing than they would in a bright office. This kind of shot helps you to create an atmosphere when doing investigative stories.

How do I do it?

The only thing you have to do is make sure your light source is very bright compared to the surroundings. Once you've arranged that, you're laughing. Sunlight streaming in through windows and ordinary household spotlights can be put to good use if you spend the time setting up the shot. Position your subject directly in the light and close the iris down as much as you dare until just their face is visible. If the light source is particularly strong you may want to close the room's other curtains or blinds to further darken the surroundings. For added effect, light a cigarette and blow the smoke around. It will look great.

If you detect this sort of tension then only one rule applies. Ask your most important question first because you may not get another chance. If the interviewee has received any media training at all then they will be sure to say exactly what they want to say first irrespective of what you ask anyway but at least you have done your duty. Anything else you get after that is a bonus.

Long interviews and noddies

We have gone on at length about how much better multi-skilled production is for certain types of television. Here's one situation where it suffers.

Recording a piece to camera on your own is a relatively simple procedure that can be done in about the same amount of time it would take a reporter and a crew. The same cannot be said of straight face-to-face interviews. I once had a producer angrily demand to know why I hadn't used a second microphone to record my questions cleanly on a different sound track. The reason was simple. I did not intend to hear my voice asking the questions. If I did then I would have had to spend ages, really ages, setting up a series of shots of myself asking the questions and nodding furiously to imagined answers.

Noddies are pictures of you, the reporter, nodding your head or reacting in some natural way to answers the interviewee is giving you. Except they are recorded after the interview has taken place. They are used as cutaways in the edit to patch together two separate items of quote when nobody remembered to change the frame after each question. They are invariably awful and rarely look 100 per cent convincing as it is difficult to get the eyeline between the reporter and interviewee entirely right. And it requires a degree of acting skill that not all reporters possess.

To film noddies, you have to set up the camera after doing the interview so that it is pointing at you, then reframe so you are looking the opposite way across the picture (or it will look like you have crossed the line). Then record a few seconds, replay it to see how you look, adjust the camera and check how you look again, and then secure your take. On a traditional shoot, the crew would just swing the camera around and tell you to get on with it.

However, there are situations where you may need to do a longer interview with somebody on your own when we see you asking the questions and nodding to the answers. If this is the case, then here's how to do it:

(1) Use two microphones and record the questions and answers on separate tracks.
(2) Position yourselves facing each other as though it was a multi-camera shoot.
(3) Record the interview reframing the subject between each answer.
(4) Make a written note of each question as you ask it.
(5) Record the interviewee performing some noddies as though they are listening to your questions.
(6) Move the camera and frame yourself.
(7) Record your intro to camera leading to the first question.
(8) Reframe and record the other questions including a 'thanks very much' at the end.
(9) Finally, for maximum flexibility, record your noddies.

Don't interrupt the interviewee when they are answering because you are limiting your editing opportunities. This interview will clearly not be presented as live and nobody is interested in your performance as a hard-hitting news journalist. They want to hear the answers. But if you follow the rules, you should have enough elements to cut together a full interview. Good luck with that one.

Set-up shots

There will always be occasions when the person you are getting a quote from has very little direct connection with the pictures in the rest of the story. You may be quoting a professional association, for instance, but all the pictures will be showing the bad work done by one of its members. You need to make sure you have some directly relevant pictures for each interview you do. These will cover the sections of the script that explain the relevance of the person who is about to pop up on the screen.

These are called set-up shots. They may show the exterior of the building, the sign on the gate, the subject walking through their office, reading a report, talking to employees. You get the idea. You need to build up at least one sequence out of these shots. For example:

CU Macintosh Enterprises sign
WS Exterior Macintosh Enterprises
MS Managing director sitting reading report
CU Managing director's face
CU The report

You can then go straight into the quote.

Including quotes in scripts

As we have already seen, quotes are incorporated into scripts like this:

SOT
IN: 10.04.56 'The reason we don't . . .'
OUT: 10.05.14 '. . . all over the place.'

You will note that we only need to write down the opening and closing words of the quote in the script rather than the entire thing. The only difficulty is deciding which quotes you are going to use and where you are going to put them. Everybody has their own method of putting scripts together. I tend to pull out the exact quotes I want at the start and then write the rest of the script around them but you don't have to do it this way. Some people write the rest of the script and choose appropriate quotes to fill the gaps.

One difference between radio scripts and television scripts is that you don't need to introduce people in the voice-over. In radio you have to include a line before the quote identifying who is about to speak. In television the picture changes, you can see it's somebody else speaking and their name and relevance to the story will pop up on screen – as long as the details are there on the left-hand side of the script.

The main points

- Know your equipment inside out to inspire confidence.
- Ensure the background has depth and life.
- Position your interviewee on one side of the picture looking across the frame.
- Record the interviewee saying and spelling their whole name and job title.
- Vary the shot from MS to CU with each answer.
- Discuss the facts in advance and devise two or three questions for the camera.
- Ask the interviewee to repeat the question with the answer.
- Position a second interviewee on the opposite side of the frame to balance the picture of talking heads within your piece.
- Remember to get a sequence of set-up shots.

11 Recording a piece to camera

Here is something that's depressing. I led a post-graduate course in television journalism and on the first day I asked which students wanted to be television presenters. A good 80 per cent put up their hands. Why is that depressing? Not because there is anything particularly wrong with wanting to be famous on the back of other people's ideas and talent but because journalism of any sort is about stories not personalities and this was most definitely a journalism course. It would be nice to think that people go into the business with at least a trace of the right motives.

That said, the piece to camera is alive and kicking and very much loved by reporters the world over. PTCs, as they are known, never add anything unique to the story. They never show us anything worthwhile. They never say anything that couldn't be said in a script. The best justification for including a PTC is that is provides yet more variety to the flow and the look of your package. That's the justification you should use when challenged.

In reality, all PTCs do is gratuitously add the face of the reporter to the package and bump up the running time by about 20 seconds. They also prove along the way that, in this day and age of television news agencies, the package was put together by a direct representative of the broadcaster.

Mind you, broadcasters are not above including a PTC into a package of pictures that they had very little to do with getting in order to personalise it in some way. It's hilarious to see a reporter traipsing off to a location that's considered appropriate to the package just to record a piece to camera. Downing Street and Parliament Square spring readily to mind for stories that have been constructed entirely out of wire copy and library pictures but evidently are thought to need that little extra injection of weight from the official political reporter. Worse, I was actually encouraged by a very senior manager at a station to find some patch of plausible woodland in London to record a PTC for a piece I had shot on location in the forests around Oslo in Norway. It was a great package. It lived without me.

Where they go

There are two types of PTC. One goes in the middle of the package. It's called a bridge. The other comes at the end and is called a signoff. As in: 'This is Phillida Fotherington-Gore for London News in Hackneyed.'

The bridge is no different to any other section of the script. Indeed, it is merely a continuation of the script. It could be covered with pictures that are a lot more relevant to the plot than the talking head of the reporter. But it's not. It's there so everybody at home can see what the reporter looks like and sleep easy at night knowing that their news was produced there at the heart of the action.

The signoff is a little more justified. It can be used to bring the story bang up to date and throw it into the future.

> 'Ministers are still discussing the motion and it looks like the debate will drag on right into the night. This is Terri Taramasalata for Worldwide News in London.'

One problem with signoffs is that so often by the time you get to them everything that could be said about a story has already been said in the voice-over so the reporter is forced to fall back on the pompous pontificating of the 'it remains to be seen' variety. It's best to avoid that. It doesn't take a great deal of original thought to come up with a new line that pushes the story on in a practical way and encourages viewers to watch a later news bulletin to see what happens.

Recording them on your own

We have already identified how difficult it is to include yourself in an interview situation because of the time it takes to frame yourself and get the eyeline right. However, it is not difficult to record a PTC at all. Like anything, you don't want to make the job more difficult than it needs to be so this is the easy way.

Find a twig or a leaf or something and put it on the spot you will be standing on. Fix the camera on the tripod and make sure the zoom is right back so you are taking the widest picture possible. This has the distinct advantage of giving your shot a very deep depth of field, guaranteeing you will be in focus. To be on the safe side, though, focus manually on the twig and then reframe with the camera horizontal so the scene in the viewfinder looks approximately right.

If the camera looks something like the one in Figure 11.1, the chances are you won't be too far out. Press record and stand on your spot for a few seconds. You can now rewind the tape and adjust the camera down or up to remove all that sky or include your head. Record another few seconds and give the tape a final check to make sure the framing is absolutely correct. If it is, you are ready for your moment of stardom.

You will be looking straight at the camera so you do not have to worry about the 'falling off the side of the picture' effect you can get with wrongly framed interviews. If there is nothing particularly interesting going on around you, you should probably stand in the middle of the frame, as in Figure 11.2. But if there is some action going on, or you are outside an unusual building, or there's just a very colourful vase of flowers lying around nearby, then you may want to stand a little to the left or to the right. Just make sure there is something in the opposite space to look at.

You may be tempted to record your PTC right by an official sign or noticeboard. There's no better way to prove you are right on top of a story about food poisoning in hospitals, for instance, than to record a PTC right by the side of the main hospital sign. But guess what? Nobody wants to read the sign. Nobody cares. It's you they

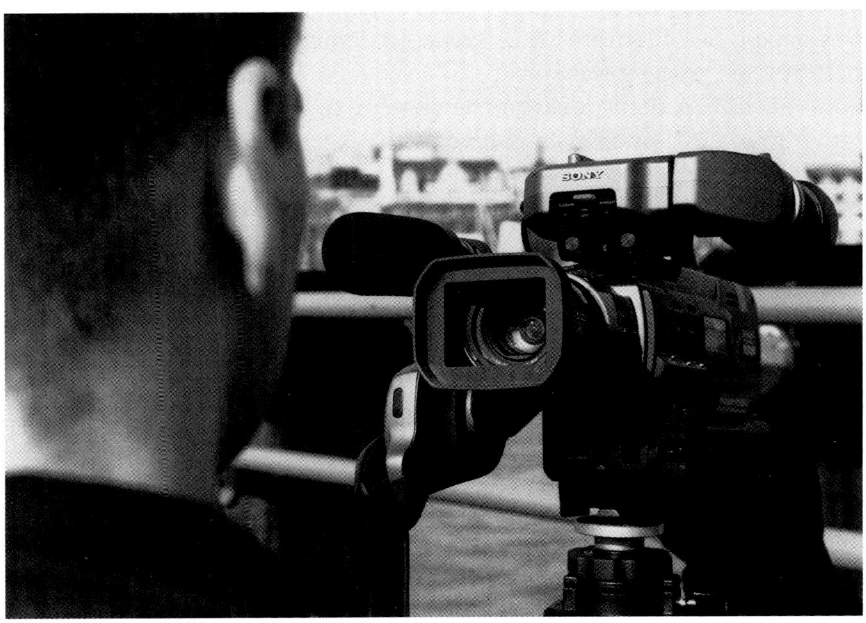

Figure 11.1
Look, Mommy, no hands . . .

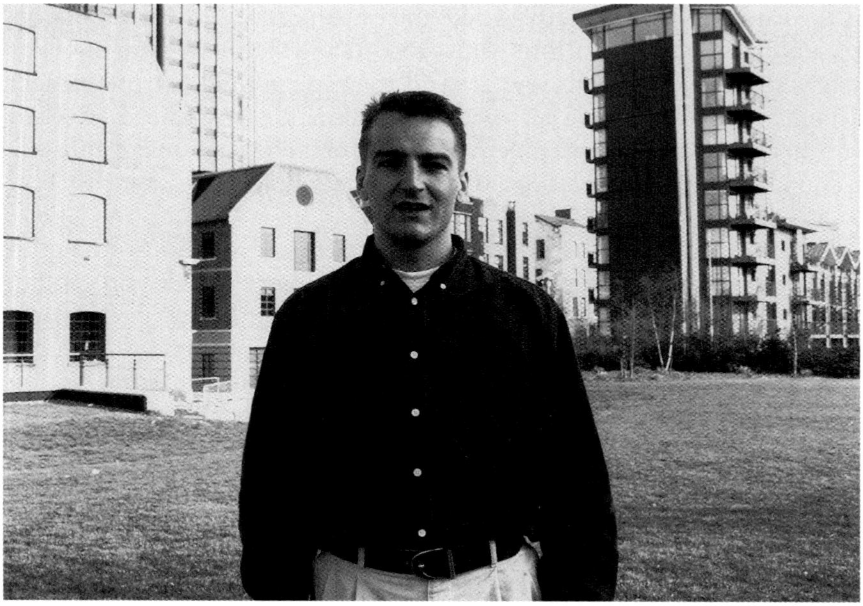

Figure 11.2
Classic framing for a PTC

are looking at. So don't sacrifice your own framing to get all the words of a sign in. By all means use the sign as a backdrop, but make sure you don't end up as a pin-head sticking up out of one corner of the screen. See Figure 11.3 to see what I mean.

If you are using a flat surface like the sign of a hospital or a wall as your backdrop then position the camera so it is looking at an angle rather than head on. This will give the background of your picture the depth you aspire to. Figures 11.4 and 11.5 show how a boring brick wall can be transformed with a little thought.

You may also have to deal with excessive backlighting. If the background is particularly bright you may find that your face ends up a dark blob in the centre of the screen. If this is the case, simply open the iris up a fraction more. The background will get even brighter but you will be correctly lit. In television terms, you would say you can live with it.

Once you are satisfied your framing is correct, you plug the hand microphone in and set the sound recording level manually. Press record, hit your spot, look straight into that lens and say something like: 'Piece to camera going in five, four, three . . .' And you count the beats for 'two' and 'one' in your head before launching into your words of wisdom. The reason 'two' and 'one' are silent is that you

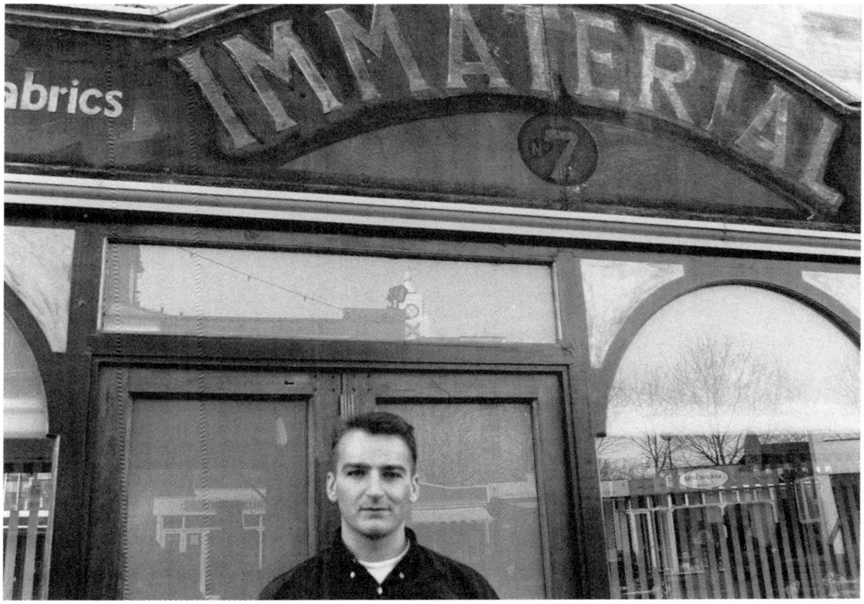

Figure 11.3
Nobody cares about the sign over your head but they will care that they can't see you properly

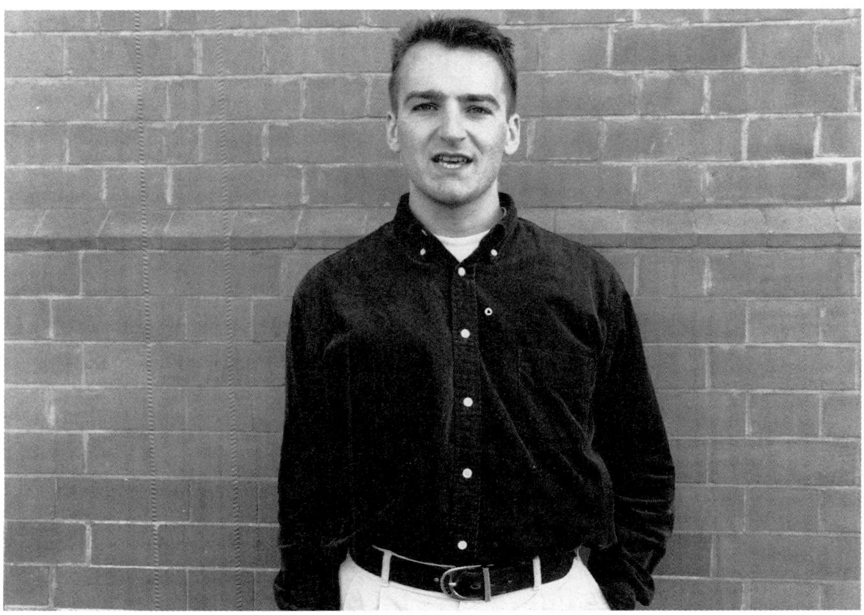

Figure 11.4
You'd have to be terribly confident that you're interesting to look at to choose this background

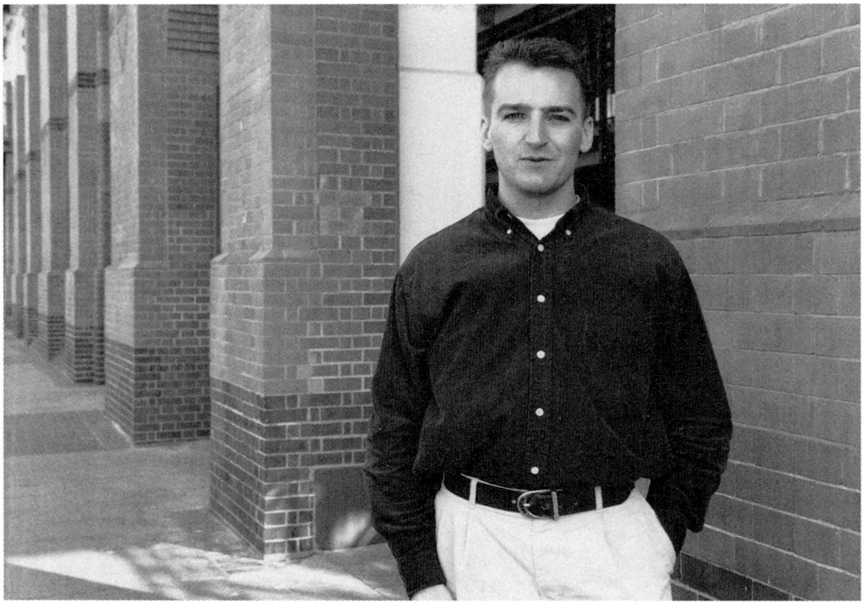

Figure 11.5
By turning the camera round a few degrees you will give your picture the depth and interest it deserves

don't want the last bit of the 'one' to roll into the first word of your PTC. It would make editing very difficult.

You may also want to identify the numerous takes you record of the same PTC as you get it wrong time and time again. You could also visually identify each take by holding up fingers just before you start. This will make it easier and quicker for you to find the good one when you are spinning through your rushes.

If you are recording a signoff, don't run away as soon as you have finished speaking. Stay there for at least ten seconds in silence looking directly into the camera. You don't need to smile falsely. This will be left on the end of your finished package so the gallery has a degree of flexibility to choose when they cut back to the studio. They will clearly only leave you up there if there's some horror going on in the studio but you standing there will look a darn sight better than a freeze frame or going to a blank screen.

Having done all that, spin the tape back and watch and listen to yourself. You should particularly listen out for aircraft, cars and dogs barking that you missed when caught up in the excitement of the actual recording. Watch out, too, for naughty little boys who run behind you making rude gestures with their hands. It happens. And it ruins your take.

The transporter effect

What is it?

Every time any station does a story that is in some way related to the television programme *Star Trek*, the stagnant and unimaginative reporter always ends it with a mocked up transporter shot around the time they do their PTC. Since Paramount holds the rights to both the actual effect and the sound that accompanies it, the reporter's version ends up rather limp and unconvincing in comparison. However, you can use the same principles in your features. For instance, somebody walks into a changing-room and half a second later walks out transformed.

How do I do it?

Put the camera on the tripod and lock it off so it cannot move. Make sure the only thing that will move within your frame is the subject. No cars, no other people, nothing. For the transporter effect, record a section without you in it. Then record yourself dematerialising and launch into your PTC. You may wish to look around in effected surprise for maximum corn. In the edit suite you simply mix between the two shots to make it seem you are appearing from nowhere. The changing-room shot is done the same way. Record the subject entering the room. Stop recording while they change. Then record them emerging. Simple.

What you should say

The biggest problem most people have when they start introducing PTCs to their packages is what to say. PTCs are recorded and set in stone well before the script is written, so how do you know it's going to fit in? The answer is: you make it up.

Like quotes from other people, you want your PTC to last around 20 seconds on screen. You know details of the story and you should have already begun to think about the structure of the finished package. All you have to do is make up a plausible section of script that does not depend on facts that you have yet to verify.

Don't write it down because you will spend all your time desperately trying to remember what it was you wanted to say. Instead, order the points in your mind and go through them one at a time. That's it. And don't, whatever you do, read it from a piece of paper taped to the side of the lens. You may think you are looking near enough to the lens but you are not. It will look odd and viewers will see your eyes moving. If the thing you want to say is so long and convoluted that you cannot remember it then prune it down to something manageable. And unless there is an overriding reason to demonstrate the fact you were present at different locations then only include one PTC per package. We've seen you once already, and that's enough.

How to look good

Standing in front of a video camera looking directly into the lens is not a particularly natural thing to do. And you can be sure that any hint of nervousness or anxiety will clearly show up. In fact it becomes a self-fulfilling prophesy. You are nervous, so you look nervous, so you are nervous. Somehow you have to break the loop.

There are two ways of learning how to speak to a camera. The first is to take a camcorder home and spend your evenings recording PTCs and reviewing the results. You can see yourself when you are getting better. Rather less elaborate is to find a mirror, look straight into your eyes and practice that way. You will be able to see exactly what is going on with your face and learn the importance of keeping your eye contact rock-steady. In their homes the viewers are looking at you. It's only courteous to look right back at them.

The main thing when recording a PTC is to keep your eye in contact with the lens throughout. If anything distracts you, the viewer will want to know what it is, even if you continue to talk.

Once when I was presenting a live news bulletin from the middle of the newsroom, someone actually came up to me and asked me a question. They stood there just off camera waiting for an answer before realising what was going on and then making the situation worse by saying 'sorry' loudly. Thankfully I managed to ignore them.

When you are recording a PTC, of course, you can afford to do it again and again until you have success. But you still have to cut out everything and everyone who is around you, including the little gaggle of interested on-lookers who are fascinated by the television-making process. That's something that comes with experience.

Something else that comes with experience is how you react to the camera. You don't want to look like your body is one big nervous twitch – nor do you want to look like a block of wood. Standing and reacting naturally to the camera is a very unnatural thing to do. You have to be very careful what your facial features do and what your hands do. Rapid movements look awful. But you don't want to descend into ham. Again, it's something that comes with experience.

You should also consider buying some makeup powder to take the shine off your face. It's not vanity. Greasy faces look awful on camera and I'm sure you don't want that. So buy some powder and at least dab it on your forehead, nose and chin just before you make your recording.

Finally, even if you find recording PTCs excruciating, you had better get used to them because you never know when you may have to do a report from the scene of a breaking news story in a live bulletin.

Putting them in scripts

Just like quotes, PTCs are identified as SOTs in the left-hand column.

PTC SOT
IN: 10.08.34 'Ministers are still . . .'
OUT: 10.08.53 '. . . News in London.'

That's it.

The main points

- Bridges go in the middle of the package.
- Signoffs come at the end.
- Perfect constant eye contact by practising PTCs in a mirror at home.
- Make the background interesting by giving it depth.
- Position yourself on the left or right of the frame if there's

something on the opposite side to balance the picture.

- Remember it's you the viewer is looking at not the sign you're standing in front of.
- Keep it to around 20 seconds and be natural.
- Check the pictures and sound you have recorded before you leave the location.
- Don't read the PTC from a note taped to the lens hood.
- Don't be frightened of using powder to stop your face shining.

12 Shooting a longer story

So far, we have concentrated on making short packages. In a lot of production environments these will be news stories and the scope for directing beautiful sequences will be somewhat limited. Indeed, time constraints will mean you simply won't be able to hang around for ages to get every last shot if you are working to a deadline. I'll always remember an irate news editor swearing that the worst thing we had been trained to do was to shoot sequences. This was a news editor who had never been trained nor worked as a videojournalist. However, it is important that you master the basics of multi-skilled television production before you go on to something bigger. And producing news packages is as good a pastime as any.

By this stage you will have shot and packaged numerous stories. You will consistently be able to come back with network-quality pictures and sound. You will be able to rattle off flowing scripts with ease combining voice-overs, quotes, PTCs and NATSOTs. You will be able to integrate the pictures with the words so they produce a satisfying whole that looks like it had some thought put into it. In short, you can walk into any location where things are happening and imagine how you would direct an effective package about it. So try it. Next time, you are in a bar or a restaurant look around you and just work out what shots you would get. In no particular order, you should think of things like:

XCU	Match being struck
CU	Match lighting cigarette
MS	Arc of smoke jetting into the air
MS	Bar tender pulling a pint
CU	Beer gushing into the pint glass
WS	Punters at the bar
XWS	Bird's eye view of the whole bar

There are many, many more shots you could get. If you have taken everything on board so far and you can do this exercise with ease everywhere you go then you can probably pat yourself on the back and start to call yourself a videojournalist. Cool, eh?

However, we now need to look at how we can use these basic skills to make slightly longer packages. Features, if you will.

Producing features

News is easy. Things happen. You film them. You ask someone what went on. You've got your package. Probably it will last between one and a half to two and a half minutes depending on how exciting the story is and what kind of pictures you have. If you find yourself getting desperate to find pictures to cover your script then you should seriously begin questioning whether the piece will make good television or not. So cut it until it works.

Features are very different. For a start, you don't generally have that absolute pressure of time. There's no hole with your name on it in a bulletin looming just two hours away. As long as the person who's in charge of your programme has a fraction of organising ability then you should always have enough time to make your piece the way you want to.

However, here again we diverge from the way conventional television has tended to be made. If you are working in a multi-skilled environment then the chances are you are using non-linear computer editing equipment. The simple machine-to-machine example we looked at in the chapter on editing is a linear edit suite. This means that once you have put something on tape, it's there and you cannot easily change it. In the days of old, producers and directors used to spend hours, days even, agonising in VHS off-line edit suites about which shots to use in what order before committing themselves to tape in the vastly more expensive on-line suites. Now computers allow us to jumble the elements around as much as we like. The only commitment comes when you dump the finished package to tape at the very end of the edit.

But this still doesn't mean you can waste lots of time doing your non-linear edit. There's no point in working efficiently on your own up to the editing stage and then descending into gross inefficiency simply because that's what has always been done.

You have learnt to structure your piece from the moment you get your first shot. You know exactly what order the pictures should go in because you shot them that way. If you are producing news to a deadline, you have an excuse not to mark your script up with every shot. There's no such get-out with features.

The trombone shot

What is it?

Movies and television drama love this one. It's used to denote any number of emotions on the part of the subject by concentrating the eye on a particular item. Imagine that you've just got home and you bound upstairs only to see your best friend come out of your bedroom in your bathrobe. The optical effect that happens to your eye at this point as you at once focus on the former best friend while everything else seems to shift perceptively out of focus is essentially what you are trying to achieve with the trombone shot.

How do I do it?

The shot is achieved by zooming in while tracking out and refocusing on the subject all at the same time. Or, for the reverse effect, zooming out while tracking in and refocusing. Your subject remains at the same point and the same size in the frame while the surroundings appear to move somehow of their own accord.

It's got to be said that this shot is easier in theory than it is in practice. If you manage it on your own you may deserve some sort of award.

The difference between news and features

News stories happen. You don't set them up. The circumstances are largely beyond your control. Features, on the other hand, allow you a lot more creative freedom both with the pictures and in the scripting.

With a feature, you are not simply reporting the events as they happened. You are perhaps profiling somebody who has an interesting hobby, or delving into the background of a current affairs story, or following the development of something over time. Anything, really, that you can get your teeth into.

Features tend to be longer than news stories and this means you have to adopt a different approach. For a start, you need more research. Knocking out a two-minute script for a news story is easy considering all the information you need will probably be right there on tap. Writing an informative, entertaining and rounded script for a four-minute feature is more involved. You need much more information about the subject.

In order to compile sufficient information talk to the people involved in great depth. Ask around the subject so you have little anecdotes to liven the proceedings up. When I went to Oslo to cover the cutting down of Trafalgar Square's Christmas tree I was told by my hosts about the trolls that lived in the forest. Apparently, they are the direct descendents of the Neanderthals. It was not immediately concerned with the main story but provided detail that let me include some beautiful shots of pine trees surrounded by snow.

As well as more information in the script, you need to get more variety in the pictures for a feature. The chances are that one location will not be enough. How many different shots can you get in a single location? How many sequences can you direct? You will soon exhaust the possibilities.

If you stay too long in the same location, your piece will begin to feel claustrophobic. So, even if your subject is essentially bound to one location you should think about other places you can go that are in some way pertinent and allow you to open the story up a bit. At the very, very least you can have, say, a person walking down the road to the shops. It will demonstrate the drudgery of their everyday lives.

You should also try to be more creative about the way you shoot your package. You need a lot of elements for a feature and you don't want to start repeating yourself. The pictures have to remain interesting throughout.

This is when your ability to frame shots in interesting ways and come up with unusual angles of perspective come to the fore. Don't make every shot bizarre because then your piece will become unwatchable. It's variety we are after and that is only limited by your imagination.

Read through the script below. It's taken from some wire copy about the birth of a new gorilla. But even that can be opened out into a news feature with lots of locations, a few interesting facts and some great pictures of cute baby animals.

(INTRO)

Next Sunday is Mother's Day so I hope you've remembered to buy a card already.

But there's one new family that won't be getting any flowers on the morning.

They're more likely to eat them or wipe their bottoms on them.

Sylvester Crystal knocked on the door.

	(BEGIN)
CU Child on swing in play area	
(NATSOT)	(NATSOT)
Children laughing	
MS Mums watching children	It's not unusual for new mums to be very protective about their offspring.
WS Play area from ground	But not many encourage them to cling round their necks.

XCU Baby gorilla's face	And even fewer have babies who weigh just five pounds and are covered in thick black hair.
(UPSOT)	(UPSOT)
Baby crying out	
MS Baby on mum's back	Meet . . . well, he doesn't have a name yet.
XWS Zoo from high point	But he was born yesterday at the National Zoo in Washington DC. He's a Lowland gorilla.
CU Zoo sign 'To the Gorillas'	And both mother and child are doing well.
(SOT)	(SOT)
IN: 10.03.04 'I was cleaning the . . .' OUT: 10.03.24 '. . . right on her back.' (IN VISION) ASTON Billy Pepperfield Zoo Keeper	'I was cleaning the chimpanzees out when I heard this sound. It was like someone was crying out for help but didn't know the words. I dropped the rake and ran over but by the time I got here all the action was over.

	Mandara was just sitting there eating the flowers. It wasn't until she turned her back that I saw this little thing with its
MS Mandara from front	arms clinging to her neck right on her back.'
XWS Mandara in cage	Mandara's the mother and she's 14 years old. It's not the first time that she's given birth.
CU Brother climbing tree	Her new son has a brother and a sister.
CU Hand gripping branch	But it will be a few days before he gets to meet the rest of his family.
MS Whoop of gorillas in tree	There are now nine members of the zoo's whoop of gorillas.
WS Gorillas through cage bars	Just like humans, they like to live in groups.
MS Baby sucking thumb	But this is one boy who's destined to come from a single parent family.

WS Bronx Zoo	His dad's called Gus and he lives way out of town.
CU Bronx Zoo sign	Gus is a silverback gorilla.
MS Gorilla cage	He's 15 years old and seems cheerfully unaware that the
CU Gus chomping on branch	product of his loins has emerged to the world.
WS Gus ripping branch apart	He's far more concerned with where the next branch is coming from.
(SOT)	(SOT)
IN: 10.12.34 'Gorillas are very . . .' OUT: 10.12.43 '. . . families I know.' ASTON Randy Bingham Animal Psychologist	'Gorillas are very maternal. It's the mother who does all the hard work and brings the children up. She breastfeeds them and carries them around until they are big enough to fend for themselves. So this little one won't miss having a father. Dads generally just sit there and eat all day. In fact, they're a lot like some human families I know.'

MS Map of gorilla areas outside cage	Lowland Gorillas usually inhabit the tropical forests in central Africa.
CU Endangered sign	But there are so few of them left in the wild that they are
CU Egg incubator	now listed as an endangered species.
MS Chicks milling around	The National Zoo is very proud of its breeding programme.
(UPSOT) Chicks squawking	(UPSOT)
	Babies are born throughout the year.
WS Vets tending to babies	A team of vets makes sure the pregnancies go as smoothly as possible.
MS Vets at work	They're the ones who mix and match males and females from other zoos.
CU Vet feeding babies	It's a bit like a dating agency. But these species depend on it.

(SOT)	(SOT)
IN: 10.26.21 'A lot of . . .'	'A lot of the animals we put
OUT: 10.26.40 '. . . the child.'	together are endangered
(IN VISION)	species. They're not like
ASTON	rabbits. They actually find it
Michael Schwartz	quite difficult to mate. Some of
Zoo Vet	them are only fertile for a few
	weeks every year so we have
	to make sure they're with a
	suitable partner at exactly the
	right time. Then we have to
WS Giraffe in own pen	take care of the mother while
	she's pregnant.'
CU Giraffe's face	Sometimes this means keeping
	the mothers isolated so there
	aren't any accidents.
MS Mandara feeding baby	But gorillas are a bit more
	hardy.
CU Mandara feeding baby	Mandara carried her boy
	without any accidents for
	eight and a half months.

(PTC SOT)	(PTC SOT)
IN: 10.28.02 'Visitors to the . . .'	'Visitors to the zoo will be able to take their own look at the mother and child from next Friday when they're both put back in the main gorilla house.
OUT: 10.28.19 '. . . National Zoo.'	
(IN VISION)	Meanwhile, the zoo keepers are keeping a close watch. They'll give the boy a name when they've seen what kind of a personality he's got. This is Sylvester Crystal for Satellite News Network at the National Zoo.'

Starting at the top, you will notice that I haven't put the timecodes in. I didn't want to clutter up the pages when there are points to be made. However, it would probably last about three minutes.

Just reading through the script, you can see all the elements unfold before your eyes. You don't need to see the rushes to envisage how this piece will look on screen. Your scripts for features should be the same.

Let's review the essential elements:

(1) Lots of locations – children's play area, baby's cage, whoop's enclosure, National Zoo exteriors, Gus's cage at Bronx Zoo, hatchery, animal surgery and the giraffe's pen.
(2) SOT from interviews – zoo keeper, animal psychologist and zoo vet.

(3) PTC – recorded in the baby's cage because it is most appropriate.
(4) NATSOT – of children playing to start it off.
(5) UPSOT – of baby and chicks making noises.
(6) IN VISION – indicating the talking head should appear just after the speech begins.
(7) ASTON – brand name for a titling machine that puts people's names on the screen.

The reporter is clearly hoping to build up a bit of mystery at the top so the intro doesn't give the game away. And, because the piece is evidently not a hard news item, the intro itself is more relaxed and informal to make the item suitable for a daytime magazine show. Things we find out about that are not directly concerned with the birth include a few facts about the status of gorillas in the wild and information on the zoo's extensive breeding programme.

So your next step should be to start making features. Again, start with something short and build up the duration and complexity. When you are totally proficient at making great features, you will have discovered the liberation and joy of being a videojournalist.

The main points

- Features give you a lot more creative control.
- Research them properly beforehand.
- Film in two or three different locations.
- Make sure you get those interesting shots for variety.
- But beware of overdoing the esoteric stuff.

13 Using your skills

Now you are a proficient camera operator, scriptwriter, producer, director and sound operator, you will want to put all those skills into action to make the kind of programming you would like to watch.

You have got many options here. If you are really good, you are only limited by what others will allow you to do.

Documentaries

Having mastered news and features, you will be choking at the bit to make a full half-hour programme. Yes! Videojournalism is entirely suitable for documentaries. In fact, it is a lot more suitable for longer programmes than it is for news. Let's look at the facts. Breaking news is a highly-pressured environment. When stories are made by crews and reporters and producers there are a lot more people around to share the different jobs and get them done in as short a time as possible. One person can set up the shoot while another works on the script and another gets the pictures. Clearly, if time is the most important factor in the production of your piece, the conventional set-up will be quicker and the people involved will be less likely to expire of a coronary. A videojournalist can produce the same news to at least the same quality but it will take longer unless he or she suddenly sprouts more arms and another head.

Documentaries, though, are another matter. I'll stick my neck out here and say, for the record, that multi-skilled techniques can produce better documentaries than conventional productions. Actually, this is not such an outrageous thing to say. For years, mad eccentrics have disappeared off to places like Russia armed only with a 16mm camera and come back to win international film awards. Nobody questions the validity of this as a method of documentary production.

Videojournalism is better for documentary production for one very simple reason:

> A director, a two-person crew, a producer, a production assistant all chattering between themselves: this is not an easy entourage to ignore.
>
> One person with a small camera: they are easy to ignore.

If your intention with the documentary is to show what is really going on as it happens, you absolutely have to fade into the background. As the film-maker, you are not the subject of the documentary. You should not dominate the proceedings. It's one thing to ask someone to press a switch after the event so you can pick up a CU shot. It's another to demand that they repeat entire conversations or actions time and time again.

Videojournalism scores the points because you can easily make yourself discrete. Your subjects won't entirely forget you are there but neither will they be constantly reminded. In addition, if you are being sensible and using small digital camcorders, your subjects will not be so intimidated by your presence. Beta cameras can be quite daunting.

Let's not look at an entire script for a half-hour programme. I'm not that desperate to bump up the pagination of this book. But there are a few points to bear in mind before embarking on your blockbuster.

Preparing a documentary

For a start, you have to make sure your subject is suitable documentary material. You do not want to commit yourself and discover half-way through the production process that there's not enough material to support a half-hour or hour-long programme.

What's the point?

You need to set up a strong narrative thread throughout its duration to keep up the interest. It may be that you can set up an issue at the start of the programme that has to be resolved by the end. Even if you cannot predict the exact outcome or the conclusion of your story, you can certainly anticipate the possible alternatives. For instance, if your documentary is about an athlete training to take part in the Olympic Games then clearly the denouement will be the contest itself – a bit like the film *Silver Dream Racer* with David Essex. If you cannot write the point of the programme down in a single sentence then maybe your idea doesn't have one and you should rethink.

Is there enough material?

Having established that your documentary has a point, you need to ensure there is enough potential material to fill it. It may help to think of your programme as a series of linked features. Again, write them down to show how these elements will pan out over the duration of the show. This is called a running order. Remember, each item in the running order must have its own point too.

What about the pictures?

Next, think about the pictures. It is vital to a successful documentary that the pictures drive the story. The pacing of documentaries is different to those of news and features. You can let the pictures breathe a little more by keeping them on the screen for longer. But even so, you don't want to use the same picture twice and you have got to make sure every one looks interesting. A half-hour documentary could need more than 250 different shots. At a bare minimum that represents around 80 different elements shot at least three different ways. And you have to get many, many times more than this to cover yourself. If you have problems envisaging this then, again, you should consider your options.

But, if everything comes together the way you anticipate in the planning stages, you are probably ready to make your documentary. How long will it take to make? If you are working on your own and everything goes according to plan, you should be able to make a rip-snorting half-hour programme comfortably in about four weeks. But it may be that your particular piece of string is longer or shorter.

Presenter-led programmes

You are also ideally placed to make programmes that are presented on location by somebody else. In this situation, you are the cameraman, sound operator, producer and director. However, you are heading for a frustrating time if the presenter needs you to tell them everything to say, feed them every question to ask and write every word of the script, as some do.

A lot of the shots of a presenter-led programme are the same ones you would get for story-led television. The WS, MS, CU shots, the cutaways, the tracking shots, the quotes, the noddies, the PTCs. You need them all. But here's where the difference comes. Presenter-led television usually includes elements that are recorded as live. This may be an interview where you will have to move the camera while you are recording from a CU of the presenter to a two-shot of them both to a CU of the interviewee and back again. Or it may be recording some sort of confrontational situation such as an embezzler being challenged on their doorstep by a leading investigative reporter.

Whichever, the hardest thing to get right in this situation is the sound. You cannot adjust the sound recording levels dynamically while you are keeping the picture correctly framed. The answer? Make sure you use the right kind of microphone and decide whether to put the levels on manual or automatic.

If both people speak about the same volume you can use a hand microphone on manual for the as-live interview. If you have no real control over the situation you will find a unidirectional microphone on automatic will be the best you can do.

Other roles you are qualified for

The skills you have learnt as a videojournalist can all be applied to conventional productions. You just have to decide which particular skill you wish to exploit. All of the following jobs make legitimate employment for the working videojournalist.

Producer

You can slot very easily into any kind of news or magazine show as a producer, exploiting your ability to research, scriptwrite and package.

Single-camera director

As a videojournalist you can make news, features or documentaries that are led either by the story or by a presenter – so you can certainly direct a two-person crew doing the same.

Camera operator

You are more than well qualified to be a news camera operator. All you have to do is learn where the buttons are on whichever camcorder it is you are given.

Presenter

If your PTCs are nice and slick, you can present programmes that are produced and directed by other people. Depending on their constitution, they may welcome your ability to write scripts, think of PTCs on the spur of the moment and come up with questions for interviewees. Presenting is an art in itself. The ability to be natural and at ease in front of the camera without appearing affected in some bizarre way is considerably rarer than you might think. The best presenters simply project an extension of their own personality. So they have to have one and they have to be able to get it across on screen. The worst get the gig because of who they think they are or what they look like (important or sexy, respectively). Still, there's an old saying that talent will win out, and if you are good then you have to believe that your day will come.

The main points

- Videojournalism is more suited to documentaries than breaking news.
- Videojournalists are much less intrusive and intimidating than a crew.
- Make sure your programme has a point.
- Write a running order to prove there is enough potential material.
- Let the pictures drive the story.
- Sound is the most difficult thing to get right on presenter-led programmes.
- Remember you can also get work as a producer, director, camera operator or presenter.

14 The multi-skilled production team

If you have been following this book from the outset, you will realise that what is being advocated is quite a revolution. Not to the people who are entering the industry – they won't know any different and will think nothing of it – but for the people who have been working in television for years.

The television industry is expanding wildly. There was once a time that Britain had only three television channels. The industry was pretty much a closed shop. There was a way of making television and it was an expensive way. But that didn't matter because the BBC had public funds coming out of its ears and the commercial station had no competition.

Not so these days. Not only are the cable, satellite and digital terrestrial channels forced to make original programming on significantly smaller budgets than their predecessors, but the competition they have brought has forced down the budgets at the traditional channels too.

Some people view these developments as anathemas. They think that the crushing of budgets leads only to de-skilling in the industry which inevitably means lower programme quality. They are wrong. It is the lack of formal training in the industry that will be its downfall not the inevitable new techniques.

All the new channels are creating jobs in television and they are all making programmes that reflect their resources. The bottom line – as I mentioned in the Preface, all those chapters ago – is does the programme get made or not? And I am one person who truly believes that it is more important to get enjoyment from making programmes and to launch channels using new techniques that reflect the resources available than to just sit around waiting for the next high-budget gig to materialise. Programmes using conventional techniques will always be made but in between them why not try something else.

If you are talented and you were well trained you will make good programmes on whatever budget you are able to secure above a certain minimum.

The right tool for the job

But does this mean that television will never be made the same way again? Of course not. There will always be the slots in the schedule and, more importantly, the budgets available to make features and documentary television using conventional methods for major series on the most popular channels. Nobody needs to feel threatened. Nobody needs to decry multi-skilling out of a need to protect their job. Outside of that narrow arena, though, informed decision-making has to come in.

The attribution of resources within lower-budgeted programmes can pretty much determine whether the show is a success or not. A low budget can result in a classy and expansive documentary using videojournalism or a sleazy cheapie that looks like it's been thrown together by monkeys at the last minute, using conventional methods.

Conversely, if you try to impose multi-skilled techniques on a programme for which they are patently unsuitable you risk ending up with unbroadcastable rubbish and no money to reshoot. At least conventional methods would have resulted in something.

Somebody who knows has to make the decision. They have to decide what kind of production techniques are going to be used on what kind of project. And in order to make that decision they have to have a firm grasp of the limitations and strengths of every kind of production methodology. At the moment, these people are rare beasts. But they are coming up through the ranks.

Deciding whether to go down this route

Whether you are setting up a new station or pondering the future of your existing department, deciding to go down the multi-skilled route is always determined by financial pressures. It's a fact of life. Let's face it: the London cable news station Channel One would not have been contemplated with the cost structure of a conventional set-up. It simply could not have the income from subscriptions and advertising to support crews, reporters, producers and the like. The alternative would have been no channel. It's better that it went on air.

Existing newsrooms and production departments have a tougher decision. Converting to multi-skilled methods is the proverbial double-edged sword. On the one hand, making television by committee using conventional methods nearly always ends up with a competent if not terribly exciting programme. It also means the people who are not very good at their jobs are carried along by the people who are good at their jobs. Crews save poor directors. Directors save poor producers. Producers save poor scriptwriters. Researchers save poor producers. And so on.

Not so with multi-skilled television. You are putting your trust and faith in individuals. The big advantage of this is that if somebody is not up to the job then there is nobody else to carry them. The pictures are unbroadcastable. The script is dire. The voice-over is bland. The solution is clear.

Then there is the question of the company structure. The cue cannot be taken from the traditional television industry. Everybody knows people in television are wildly overpaid and there are great riches to be had – on the conventional channels, that is. But the corporate structure of a multi-skilled environment should, again, reflect the resources available. If a station is the television equivalent of the local newspaper, employing videojournalists instead of reporters, it doesn't need to adopt a cost structure based on the industry's going rate. And that means from the editor-in-chief down.

However, if you are setting up a documentary department based on multi-skilled techniques producing programmes for major broadcasters, you expect and indeed need to employ videojournalists of the highest calibre – and these people cannot be paid peanuts or they will go elsewhere. But if you are employing the industry's best then you have to be aware of what you have employed: a highly-trained professional journalist who is more than capable of producing, directing, scriptwriting and shooting the socks off the best of them. You do not employ people like this and sit on them. They will leave.

Going for it

So if you are considering converting to multi-skilling, you have to be prepared to rethink the whole ethos of making television from a different angle. This will be easy for the open-minded and those entering the industry and difficult for the stubborn and bloody-minded.

People have to embrace the concept of videojournalism and see for themselves the many advantages of being multi-skilled. You do not want a department full of people who constantly hark back to the 'good old days' when all they had to do was turn up and record a piece to camera. The good old days are now. They have been liberated.

The approach of the staff will largely be determined by the approach of the management and the way multi-skilling is explained. For instance:

> Approach one: 'Get multi-skilled or lose your jobs. The money's running out.'
>
> Approach two: 'Let's have your ideas for features and documentary programmes. We've found a way we can afford to make them.'

It doesn't take a genius to work out which approach will result in the most amenable and productive workforce.

Setting up a multi-skilled production environment

There are no hard and fast rules for establishing a multi-skilled production office. There are too many variables based on the budgets available, the kind of programming required and the number of hours to be filled. However, there is one thing that every multi-skilled department has in common and it is something that uniquely distinguishes videojournalism from other forms of production. Everyone in the multi-skilled department has a bona fide answer to the question: what do you actually do here? And they can physically show you the results of their efforts.

A brand new department

If you are setting up a brand new production office based on multi-skilled production techniques you have the easiest job. At a bare minimum, newsrooms can be set up in exactly the same way as local newspapers are run.

The multi-skilled newsroom

The news editor manages the videojournalists and decides on the mix of stories. He or she assigns jobs from the forward planning department and gives final approval to the scripts. It needs to be no more hierarchical or complicated than that.

When journalists come up with a story, they liaise with both the news editor to make sure the piece is actually wanted and the forward planning department to find an appropriate slot for it to go in. This way, the journalists get to be the journalists they were trained to be. And the station gets its airtime filled with suitable packages.

Features production

A feature programme can be run pretty much the same way. You need one senior videojournalist in charge of the whole show and then a team of others who go out and make the features. Instead of having a daily news meeting you have a weekly production meeting. You would probably need between four and six people working on a weekly features programme.

Documentary department

Documentaries can be made by individual videojournalists who liaise with a single series producer. Give them about a month on their own and they will come up with a great half-hour programme. Or you have a senior videojournalist and an assistant making the programme.

Converting from a conventional setup

Before you set off on the multi-skilled route, you have to ask yourself some fundamental questions about what you are trying to achieve and how you expect to achieve it. You don't want to trash your current setup unless you have something bigger and better to replace it. Thankfully, videojournalism can be a better option and it can embrace producers, directors, researchers, crews and reporters if it is properly handled.

We can assume that the overriding motivation to convert is a desire to cut costs. But that in itself is not enough. Remember, bigger and better. And cutting costs on its own is merely downsizing. You can do much more:

- Increase the quality of your programming.
- Broadcast more documentaries in your schedule.
- Increase the number of stories in your news bulletins.
- Widen the breadth of your coverage.

And you can still manage to cut your production budget so the accountants are happy. So the kind of questions you need to ask yourself are:

(1) What kind of programming do I want to make?
(2) How many people can I afford to employ?
(3) How much equipment can I afford to buy?

If you want to make more documentaries and feature programmes, then great. If you simply want to cover the latest breaking news then consider a mix of multi-skilled people and conventional crews. And put somebody in charge of them who understands which employees are suitable for which stories.

The big-bang approach

Let's say you produce daily bulletins that concentrate on both news and features. You think it would be ideal to incorporate the multi-skilled approach and you are probably right. You know the current structure and you have the eventual structure on paper. But somewhere in between there has to be a training programme. The problem you have to sort out is how to keep making news items while you train people.

One solution could be to split your team into two and train one lot of people first while the others continue to produce the bulletins (they should view it in the same way as the middle of summer when everyone else is on holiday) and then train them as the first lot implement the new structure. That would work if it was correctly handled.

The new-department approach

An alternative approach would be to establish a completely new department within your existing structure dedicated to multi-skilled production techniques. First, you could ask for volunteers from your existing workforce to undergo the training programme. They would form the core of your new department. To begin with, give them just one programme to produce. Then gradually, over time, hand more of your airtime over to your new department. More people can then be trained up and your conventional setup will shrink.

This approach has the added advantage of allowing you to monitor the success of the multi-skilled approach without any critical commitment. It gives you time to discover the strengths and weaknesses for yourself, and to work out the best way to use videojournalism for your own purposes.

The bit-by-bit approach

The truly timid will prefer to dip their toes in the water first. Maybe you just retrain one local bureau to see how it goes. Again, back-to-basics training is everything. Otherwise your brave experiment will end in failure.

The only thing you have to bear in mind by adopting this approach is the reactions of your existing employees to the prodigy who is seemingly able to produce top-quality television on their own. This should be anticipated and dealt with in some way.

Health and safety

Traditionally, editorial people have been in charge of journalists and technical people have been in charge of crews. The technical people know there are courses that crews go on to learn about health and safety in the workplace – from making sure they don't electrocute themselves in heavy rain to preventing third parties from tripping over sound cables to ensuring they get out of riot situations and scrums alive and in one piece. The technical people know all about preventing accidents from happening. Not so the editorial people. That side of things has always been taken care of by the crews so they have been largely ignorant of it. But as soon as you start multi-skilling journalists you really have to start becoming aware of it. And the videojournalists have to be taught to the same standard that the technical people have always been.

But the issue of health and safety for multi-skilled people goes a lot further than that. There's a story of a lone videojournalist who was sent on an assignment into the roughest most notorious housing estate in the area. He returned saying how he had to face off a gang who were trying to break into his car. He had to face off this gang while he was carrying about £15,000 worth of camera equipment. This was no laughing matter. And it was something that nobody should have to face alone in the course of what is only a job. But he had been ordered to go there by somebody who didn't think.

So as well as ensuring that the videojournalists learn all about health and safety, the people who are managing the operation have to be brought up to speed about the needs and requirements of the multi-skilled workforce.

One final note

So now you know everything there is to know about multi-skilled television production. You can see how it works and why it works. And, just as importantly, you can see when it shouldn't be used.

Before I take my leave, there's just one more thing to share. I don't claim responsibility for it. I found it hanging up outside the boss's office at the breakfast station GMTV. It's a list called the six stages of television production. Not all productions and stations, of course, are like this. Just a lot of them. Your job is to make sure any project you are in charge of is not one of them.

The six stages of production

(1) Wild enthusiasm.
(2) Total confusion.
(3) Utter despair.
(4) Search for the guilty.
(5) Persecution of the innocent.
(6) Promotion of the incompetent.

It's so true. Oh so horribly true.

Glossary

As live Term meaning that the recorded piece, usually fronted by a presenter on location, will not be edited before broadcast.

Beta SP Established professional broadcast quality tape format that has pretty much become the industry standard.

Colour bars A colour test signal generated by some camcorders that looks like a testcard.

CU Close Up – shot, such as a flower head or smoking cigarette, that gives your piece essential visual detail.

Cut aways Shots that allow you to cut away from the main action to compact time or cover directing mistakes.

DV Digital videotape format designed to replace Hi-8 but just about good enough for broadcast.

Editing

Linear Conventional editing, where your finished package is recorded on tape as you go along.

Non-linear Computer-based editing, where a piece is stored on a hard disk, allowing a lot more flexibility to rearrange shots.

Off-line Cheap, low-quality editing, often using domestic VHS machines, that results in a rough cut.

On-line Expensive, broadcast-quality editing that results in the final cut.

Hi-8 Semi-professional tape format that is almost but not quite broadcast-quality and has now been superseded by DV.

Intro The script read by the presenter that introduces a package.

IN VISION Term marked on the script that determines when the source of the SOT appears on screen.

Line, The An axis through every picture which you must not cross.

Microphone

Clip Tiny omnidirectional microphone clipped to a loose flap of the interviewee's clothing. Also called a tie microphone.

Hand Omnidirectional microphone used for PTCs, location interviews and karaoke sessions.

Omnidirectional Microphone that picks up sound from every direction.

Unidirectional Microphone that only picks up sound from the direction it's pointing in.

MS Medium Shot – one that is framed halfway between a CU and a WS.

NATSOT NATural Sound On Tape – term marked on script to indicate which of the background sounds should be heard.

Noddies Cutaways of interviewers or interviewees reacting with a variety of expressions to answers or questions.

NTSC National Television Standards Committee – the US television standard of 30 frames per second.

Package Collective noun for all the elements of a finished piece including the script, voice-over, rushes and perhaps graphics.

PAL Phase Alteration by Line – common European television standard of 25 frames per second.

Pan A camera shot that moves horizontally from one important item to another.

PTC Piece To Camera – a section of script recorded in vision by the reporter on location.

Rushes Unedited footage straight from the camera.

SECAM Sequential Couleur A Mémoire – television standard of 25 frames per second used, for instance, in France.

SOT Sound On Tape – mark on the script indicating that the sound on tape, usually a quote from an interviewee, should be used.

Tilt A camera shot that moves vertically from the ground, usually used to denote how high buildings are.

Timecode Counter that uniquely identifies every sequential frame on the tape.

Tracking shot Pan-like shot which follows a moving object.

UPSOT Turn UP the Sound On Tape – mark on script indicating that the background noise should be brought to the fore.

Wildtrack About one minute of background sound recorded on location to help smooth picture edits.

WS Wide Shot – usually establishing where everything is in relation to everything else or just how spectacular the vista really is.

XCU eXtreme Close Up.

XWS eXtreme Wide Shot.

Index